π

Advance praise

A Biography of the World's Most Mysterious Number

Alfred S. Posamentier
& Ingmar Lehmann

Afterword by Dr. Herbert A. Hauptman,
Nobel Laureate

 Prometheus Books

59 John Glenn Drive
Amherst, New York 14228-2197

Published 2004 by Prometheus Books

Inquiries should be addressed to
Prometheus Books
59 John Glenn Drive
Amherst, New York 14228–2197
VOICE: 716–691–0133, ext. 207
FAX: 716–564–2711
WWW.PROMETHEUSBOOKS.COM

08 07 06 05 04   5 4 3 2 1

Library of Congress Cataloging-in-Publication Data

Posamentier, Alfred S.
    Pi : a biography of the world's most mysterious number / Alfred S. Posamentier and Ingmar Lehmann.
        p. cm.
    Includes bibliographical references and index.
    ISBN 1–59102–200–2 (hardcover : alk. paper)
    1. Pi. I. Lehmann, Ingmar. II. Title.

QA484.P67 2004
512.7'3—dc22

2004009958

Printed in the United States of America on acid-free paper

Contents

6 Contents

Acknowledgments

The daunting task of describing the story of π for the general reader had us spend much time researching and refreshing the many tidbits of this fascinating number that we encountered in our many decades engaged with mathematics. It was fun and enriching. Yet the most difficult part was to be able to present the story of π in such a way that the general reader would be able to share the wonders of this number with us. Therefore, it was necessary to solicit outside opinions. We wish to thank Jacob Cohen and Edward Wall, colleagues at the City College of New York, for their sensitive reading of the entire manuscript and for making valuable comments in our effort to reach the general reader. Linda Greenspan Regan, who initially urged us to write this book, did a fine job in critiquing the manuscript from the viewpoint of a general audience. Dr. Ingmar Lehmann acknowledges the occasional support of Kristan Vincent in helping him identify the right English words to best express his ideas. Dr. Herbert A.

Hauptman wishes to thank Deanna M. Hefner for typing the afterword and Melda Tugac for providing some of the accompanying figures. Special thanks is due to Peggy Deemer for her marvelous copyediting and for apprising us of the latest conventions of our English language while maintaining the mathematical integrity of the manuscript.

It goes without saying that the patience shown by Barbara and Sabine during the writing of this book was crucial to its successful completion.

Preface

Surely the title makes it clear that this is a book about π, but you may be wondering how a book could be written about just one number. We will hope to convince you throughout this book that π is no ordinary number. Rather, it is special and comes up in the most unexpected places. You will also find how useful this number is throughout mathematics. We hope to present π to you in a very "reader-friendly" way—mindful of the beauty that is inherent in the study of this most important number.

You may remember that in the school curriculum the value that π took on was either 3.14, $3\frac{1}{7}$, or $\frac{22}{7}$. For a student's purposes, this was more than adequate. It might have even been easier to simply use $\pi = 3$. But what is π? What is the real value of π? How do we determine the value of π? How was it calculated in ancient times? How can the value be found today using the most modern technology? How might π be used? These are just some of the questions that we will explore as you embark on the chapters of this book.

We will begin our introduction of π by telling you what it is and roughly where it came from. Just as with any biography (and this book is no exception), we will tell you who named it and why, and how it grew up to be what it is today. The first chapter tells you what π essentially is and how it achieved its current prominence.

In chapter 2 we will take you through a brief history of the evolution of π. This history goes back about four thousand years. To understand how old the concept of π is, compare it to our number system, the place value decimal system, that has only been used in the Western world for the past 802 years![1] We will recall the discovery of the π ratio as a constant and the many efforts to determine its value. Along the way we will consider such diverse questions as the value of π as it is mentioned in the Bible and its value in connection with the field of probability. Once the computer enters the chase for finding the "exact" value of π, the story changes its complexion. Now it is no longer a question of finding the mathematical solution, but rather how fast and how accurate can the computer be in giving us an ever-greater accuracy for the value of π.

Now that we have reviewed the history of the development of the value of π, chapter 3 provides a variety of methods for arriving at its value. We have chosen a wide variety of methods, some precise, some experimental, and some just good guessing. They have been selected so that the average reader can not only understand them but also independently apply them to generate the value of π. There are many very sophisticated methods to generate the value of π that are well beyond the scope of this book. We have the general reader in mind with the book's level of difficulty.

---

1. The first publication in western Europe, where the Arabic numerals appeared, was Fibonacci's book *Liber abaci* in 1202.

With all this excitement through the ages centered on π, it is no wonder that it has elicited a cultlike following in pursuit of this evasive number. Chapter 4 centers on activities and findings by mathematicians and math hobbyists who have explored the value of π and related fields in ways that the ancient mathematicians would never have dreamed of. Furthermore, with the advent of the computer, they have found new avenues to explore. We will look at some of these here.

As an offshoot of chapter 4, we have a number of curious phenomena that focus on the value and concept of π. Chapter 5 exhibits some of these curiosities. Here we investigate how π relates to other famous numbers and to other seemingly unrelated concepts such as continued fractions. Again, we have limited our presentation to material that would require no more mathematical knowledge than that of high school mathematics. Not only will you be amused by some of the π equivalents, but you may even be inspired to develop your own versions of them.

Chapter 6 is dedicated to applications of π. We begin this chapter with a discussion of another figure that is very closely related to the circle but isn't round. This Reuleaux triangle is truly a fascinating example of how π just gets around to geometry beyond the circle. From here we move on to some circle applications. You will see how π is quite ubiquitous—it always comes up! There are some useful problem-solving techniques incorporated into this chapter that will allow you to look at an ordinary situation from a very different point of view—which may prove quite fruitful.

In our final chapter, we present some astonishing relationships involving π and circles. The situation that we will present regarding a rope placed around the earth will surely challenge everyone's intuition. Though a relatively short chapter, it will surely surprise you.

It is our intention to make the general reader aware of the myriad of topics surrounding π that contribute to making mathematics beautiful. We have provided a bibliography of this famous number and many of its escapades through the fields of mathematics. Perhaps you will feel motivated to pursue some of these aspects of π further, and some of you may even join the ranks of the π enthusiasts.

Alfred S. Posamentier
and Ingmar Lehmann
April 18, 2004

Chapter 1

What Is π?

Introduction to π

This is a book about the mysterious number we call π (pronounced "pie," while in much of Europe it is pronounced "pee"). What most people recall about π is that it was often mentioned in school mathematics. Conversely, one of the first things that comes to mind, when asked what we learned in mathematics during our school years, is something about π. We usually remember the popular formulas attached to π, such as $2\pi r$ or πr^2. (To this day, there are adults who love to repeat the silly response to πr^2: "No, pie are round!"). But do we remember what these formulas represent or what this thing called π is? Usually not. Why, then, write a book about π? It just so happens that there is almost a cultlike following that has arisen over the concept of π. Other books have been written about π. Internet Web sites report about its "sightings," clubs meet to discuss its properties, and even a day on the calendar is set aside to celebrate it, this being March 14, which coincidentally just happens also to be Albert Einstein's birthday (in 1879). You may be wondering how March 14 was selected as π day. For those who remember the common value (3.14) that π took on in the schools, the answer will become obvious.[1]

It surely comes as no surprise that the symbol π is merely a letter in the Greek alphabet. While there is nothing special about this particular letter in the Greek alphabet, it was chosen, for reasons that we will explore later, to represent a ratio that harbors curious intrigue and stories of all kinds. It found its way from a member of the Greek alphabet to represent a most important geometric constant and subsequently has unexpectedly appeared in a variety of other areas of mathematics. It has puzzled generations of mathematicians who have been challenged to define it, determine its value, and explain the many related areas in which it sometimes astoundingly appears. Ubiquitous numbers, such as π, make mathematics the interesting and beautiful subject that many find it to be. It is our intent to demonstrate this beauty through an acquaintance with π.

Aspects of π

Our aim here is not to decipher numerous complicated equations, to solve difficult problems, or to try to explain the unexplainable. Rather, it is to explore the beauty and even playfulness of this famous number, π, and to show why it has inspired centuries of mathematicians and math enthusiasts to further pursue and investigate its related concepts. We will see how π takes on unexpected roles, comes up in the most unexpected places, and provides the

---

1. In the United States we write the date as 3/14.

never-ending challenge to computer specialists of finding ever-more-accurate decimal approximations for the value of π. Attempts at getting further accuracy of the value of π may at first seem senseless. But allow yourselves to be open to the challenges that have intrigued generations of enthusiasts.

The theme of this book is understanding π and some of its most beautiful aspects. So we should begin our discussion and exploration of π by defining it. While for some people π is nothing more than a touch of the button on a calculator, where then a particular number appears on the readout, for others this number holds an unimaginable fascination. Depending on size of the calculator's display, the number shown will be

3.1415927,
3.141592654,
3.14159265359,
3.14159265358979323846264433832795, or even longer.

This push of a button still doesn't tell us what π actually is. We merely have a slick way of getting the decimal value of π. Perhaps this is all students need to know about π: that it represents a specific number that might be useful to know. However, here students would be making a colossal mistake to dismiss the importance of the topic, by just focusing on the application of π in particular formulas and getting its value automatically just by the push of a button.

The Symbol π

The symbol π is the sixteenth letter of the Greek alphabet, yet it has gained fame because of its designation in mathematics. In the Hebrew and the Greek languages of antiquity, there were no numerical symbols. Hence, the letters of the respective alphabets served as numerical symbols. Since the Greek alphabet had only twenty-four letters, though twenty-seven were needed, they used three letters of Semitic origin, namely, Ϝ [digamma] (for 6), Ϙ [qoph] (for 90), and ϡ [san] (for 900).

The Greeks at the beginning of the fifth century BCE then used the notation represented in the following table:[2]

α	β	γ	δ	ε (, ϵ)	Ϝ	ζ	η	θ
1	2	3	4	5	6	7	8	9
ι	κ	λ	μ	ν	ξ	ο	π	Ϙ
10	20	30	40	50	60	70	**80**	90
ρ	σ	τ	υ	φ	χ	ψ	ω	ϡ
100	200	300	400	500	600	700	800	900
,α	,β	,γ	,δ	,ε (, ϵ)	,Ϝ	,ζ	,η	,θ
1,000	2,000	3,000	4,000	5,000	6,000	7,000	8,000	9,000

Thus in the old Greek texts π was used to represent the number 80. By coincidence, the Hebrew letter פ (pe) has the same value.

Recollections of π

Perhaps by coincidence or by some very loose associations, the letter π was later chosen by mathematicians to represent a very important constant value related to the circle. Remember, the circle is the most symmetric plane geometric figure and one that goes back in history to prehistoric times. Specifically, π was chosen to represent *the ratio*

---

2. A comma at the left indicates thousands. The ten thousands are indicated with an **M** below the number symbol. Table from Georges Ifrah, *Universal History of Numerals* (New York: Campus, 1986), p. 289.

of the circumference of a circle to its diameter.[3] This would be expressed symbolically as $\pi = \frac{C}{d}$, where C represents the length of the circumference and d represents the length of the diameter. The diameter of a circle is twice the length of the radius, $d = 2r$, where r is the length of the radius. If we substitute $2r$ for d, we get $\pi = \frac{C}{2r}$, which leads us to the famous formula for the circumference of a circle: $C = 2\pi r$, an alternative of which is $C = \pi d$.

The other familiar formula containing π is that the area of a circle is πr^2. This formula is more complicated to establish than that for the circumference of the circle, which followed directly from the definition of π.

Formula for the Area of a Circle

Let's consider a relatively simple "derivation" for the formula ($A = \pi r^2$) for the area of a circle with radius r. We begin by drawing a convenient-size circle on a piece of cardboard. Divide the circle (which consists of $360°$) into sixteen equal arcs. This may be done by marking off consecutive arcs of $22.5°$ or by consecutively dividing the circle into two parts, then four parts, then bisecting each of these quarter arcs, and so on.

---

3. A purist might ask: how do we know that this ratio is the same for all circles? We will assume this constancy for now.

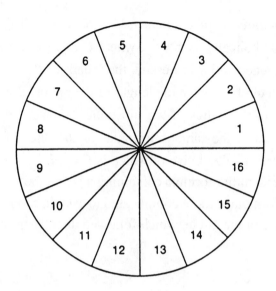

Fig. 1-1

The sixteen sectors we have constructed (shown above) are then to be cut apart and placed in the manner shown in the figure below.

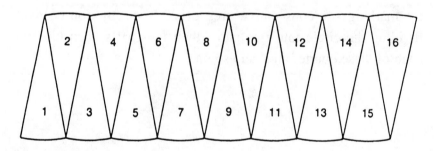

Fig. 1-2

This placement suggests that we have a figure that approximates a parallelogram.[4] That is, were the circle cut into more sectors, then the figure would look even more like a true parallelogram. Let us assume it is a parallelogram. In this case, the base would have a length of half the circumference of the original circle, since half of the circle's arcs are used for each of the two sides of the approximate parallelogram. In other words, we formed something that resembles a parallelogram where one pair of opposite sides are not straight lines, rather they are circle arcs. We will progress as though they were straight lines, realizing that we will have lost some accuracy in the process. The length of the base is $\frac{1}{2}C$. Since $C = 2\pi r$, the base length is, therefore, πr. The area of a parallelogram is equal to the product of its base and altitude. Here the altitude is actually the radius, r, of the original circle. Therefore, the area of the "parallelogram" (which is actually the area of the circle we just cut apart) is $(\pi r)(r) = \pi r^2$, which gives us the commonly known formula for the area of a circle. For some readers this might be the first time that the famous formula for the area of a circle, $A = \pi r^2$, actually has some real meaning.

The Square and the Circle

Without taking the reader's attention too far afield, it might also be interesting to point out that π has the unique distinction of taking the area of a square, whose side has the length of the radius of a circle, and converting its area to that of the circle. It is the constant value connector in this case. The area of the square (fig. 1-3) is r^2 and, when multiplied by π, gives us the area of the circle: πr^2.

---

4. A parallelogram is a quadrilateral (a four-sided polygon) with opposite sides parallel.

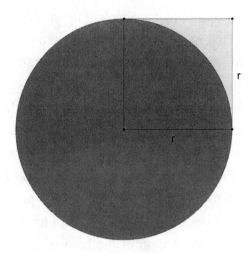

Fig. 1-3

The Value of π

Now that we have an understanding of what π meant in the context of these old familiar formulas, we shall explore what the actual value is of this ratio π. One way to determine this ratio would be to carefully measure the circumference of a circle and its diameter and then find the quotient of these two values. This might be done with a tape measure or with a piece of string. An extraordinarily careful measurement might yield 3.14, but such accuracy is rare. As a matter of fact, to exhibit the difficulty of getting this two-place accuracy, imagine twenty-five people carrying out this measurement experiment with different-size circular objects. Imagine then taking the average of their results (i.e., each of their measured circumferences divided by their measured diameters). You would likely be hard pressed to achieve the accuracy of 3.14.

You may recall that in school the commonly used value for π is 3.14 or $\frac{22}{7}$. Either is only an approximation. We cannot get the exact value of π. So how does one get a value for π? We will now look at some of the many ingenious ways that mathematicians over the centuries have tried to get ever-more-precise values for π. Some are amusing; others are baffling. Yet most had significance beyond just getting closer approximations of π.

One of the more recent attempts to get a closer approximation of π took place in Tokyo. In his latest effort, in December 2002, Professor Yasumasa Kanada (a longtime pursuer of π) and nine others at the Information Technology Center at Tokyo University calculated the value of π to 1.24 trillion decimal places, which is six times the previously known accuracy, calculated in 1999. They accomplished this feat with a Hitachi SR8000 supercomputer, which is capable of doing 2 trillion calculations per second. You may ask, why do we need such accuracy for the value of π? We don't. The methods of calculation are simply used to check the accuracy of the computer and the sophistication of the calculating procedure (sometimes referred to as an algorithm), that is, how accurate and efficient it is. Another way of looking at this is how long will it take the computer to get an accurate result? In the case of Dr. Kanada, it took his computer over six hundred hours to do this record-setting computation.

It might be worthwhile to consider the magnitude of 1.24 trillion. How old do you think a person who has lived 1.24 trillion seconds might be? The question may seem irksome since it requires having to consider a very small unit a very large number of times. However, we know how long a second is. But how big is one trillion? A trillion is 1,000,000,000,000, or one thousand billion. Thus, to calculate how many seconds there are in one year: $365 \times 24 \times 60 \times 60 = 31,536,000$ seconds. Therefore, $\frac{1,000,000,000,000}{31,536,000} = 31,709.79198376458650431253170979 \approx 31,710$ years, or one would have to be in his 31,710th year of life to have lived one trillion seconds!

The value of π continues to fascinate us. Whereas a common fraction results in a periodic decimal, π does not. A periodic decimal is a decimal that eventually repeats its digits indefinitely. Consider the common fraction $\frac{1}{3}$. By dividing 1 by 3, we get its decimal equivalent as 0.3333333.[5] This decimal has a period of one, which means that the one digit, 3, repeats indefinitely. Here are some other periodic decimals:

$$\frac{1}{2} = .5000\overline{0}, \ \frac{2}{3} = .666\overline{6}, \text{ and } \frac{2}{7} = 0.285714285714\overline{285714}.$$

We place a bar over the last repeating period to indicate its continuous repetition. The decimal $\frac{2}{7}$ has a period of six, since there are six places continuously repeating.

There is no periodic repetition in the decimal value of π. As a matter of fact, although some would use the decimal approximation of π to many places as a table of random numbers—useful in randomizing a statistical sample—there is even a flaw there. When you look at, say, the first 1,000 decimal places of π, you will not see the same number of each of the ten digits represented. Should you choose to count, you will find that the digits do not appear with equal frequency even in the first 150 places. For example, there are fewer sevens (10 in the first 150 places) than threes (16 in the first 150 places). We will examine this situation later.

π Peculiarities

There are many peculiarities in this list of digits. Mathematician John Conway has indicated that if you separate the decimal value

---

5. The bar over the 3 indicates that the 3 repeats indefinitely.

of π into groups of ten places, the probability of each of the ten digits appearing in any of these blocks is about one in forty thousand. Yet he shows that it does occur in the seventh such group of ten places, as you can see from the grouping below:

π = 3.1415926535   8979323846   2643383279   5028841971
6939937510 5820974944 $\boxed{5923078164}$ 0628620899 8628034825
3421170679 8214808651 3282306647 0938446095 5058223172
5359408128. . .

Another way of saying this is that every other grouping of ten has at least one repeating digit. The sums of these digits also show some nice results: the sum of the first 144 places is 666, a number with some curious properties as we shall see later.

On occasion, we stumble upon phenomena involving π that have nothing whatsoever to do with a circle. For example, the probability that a randomly selected integer (whole number) has only unique prime divisors[6] is $\frac{6}{\pi^2}$. Clearly this relationship has nothing to do with a circle, yet it involves the circle's ratio, π. This is just another feature that adds to the centuries-old fascination with π.

The Evolution of the Value of π

There is much to be said for the adventures of calculating the value of π. We will consider some unusual efforts in the next few chapters. However, it is interesting to note that Archimedes of Syracuse

---

6. "Unique prime divisors" refers to divisors of a number that are prime numbers and not used more than once. For example, the number 105 is a number with unique prime divisors: 3, 5, and 7, while 315 is a number that does not have unique prime divisors: 3, 3, 5, and 7, since the prime divisor 3 is repeated.

(287–212 BCE) showed the value of π to lie between $3\frac{10}{71}$ and $3\frac{1}{7}$. That is,

$$3\frac{10}{71} < \pi < 3\frac{1}{7}$$

$$\frac{223}{71} < \pi < \frac{22}{7}$$

$$3.1408\ldots < \pi < 3.1428\ldots$$

The Dutch mathematician Ludolph van Ceulen (1540–1610) calculated π to thirty-five places, so for a time the ratio π was called *Ludolph's number*. When Ludolph van Ceulen finished his calculations, he wrote the following: "Die lust heeft, can naerder comen" ("The one who has the desire, can come closer").

Another early technique for calculating π was discovered by John Wallis (1616–1703), a professor of mathematics at Cambridge and Oxford universities, who subsequently published it in his book, *Arithmetica infinitorum* (1655). There he presented a formula for π (actually $\frac{\pi}{2}$, which we then merely double to get π). The following is Wallis's formula:

$$\frac{\pi}{2} = \frac{2\times2}{1\times3} \times \frac{4\times4}{3\times5} \times \frac{6\times6}{5\times7} \times \frac{8\times8}{7\times9} \times \cdots \times \frac{2n\times2n}{(2n-1)\times(2n+1)} \times \cdots$$

This product converges to the value of $\frac{\pi}{2}$. That means it gets closer and closer to the value of $\frac{\pi}{2}$ as the number of terms increases.

What is it about the value of π that evokes so much fascination? For one, it cannot be calculated by a combination of the operations of addition, subtraction, multiplication, and division, which was suspected by Aristotle (384–322 BCE). He hypothesized that π is an irrational number;[7] in other words, the circumference and the radius of a circle are incommensurable. That means there doesn't exist a common unit of measure that will allow us to measure both the circumference and the radius. This was proved in 1806[8] by the French mathematician Adrien-Marie Legendre (1752–1833)—more than two millennia later!

But even more fascinating is the fact that π cannot be calculated by a combination of the operations of addition, subtraction, multiplication, division, *and square root extraction*. This means π is a type of nonrational number called a transcendental number.[9] This was already suspected by the Swiss mathematician Leonhard Euler (1707–1783),[10] but it was first proved in 1882 by the German mathematician (Carl Louis) Ferdinand Lindemann (1852–1939). Remember, it is sometimes more difficult to prove that something cannot be done than to prove it is possible to be done. Thus, for Lindemann to establish that π could not be produced by a combination of the five operations—addition, subtraction, multiplication, division, and square root extraction—was quite an important contribution to the development of our understanding of mathematics.

The establishment of the transcendence of π extinguished the hopes of all those who sought a method to "square the circle," that is, to construct[11] a square of side *s*, such that its area equals that of the given circle of radius *r*. Lindemann killed that hope for all time.

---

7. An irrational number is one that cannot be expressed as a fraction that has integers in its numerator and denominator.

8. The proof in 1767 by the German mathematician Johann Heinrich Lambert (1728–1777) had a flaw in it.

9. A transcendental number is one that is not the root of a polynomial equation with rational coefficients. Another way of saying this is that it is a number that cannot be expressed as a combination of the four basic arithmetic operations and root extraction. In other words, it is a number that cannot be expressed algebraically. π is such a number.

10. The term *transcendental number* was introduced by Euler.

11. By "construct" we refer to the Euclidean constructions, namely, using a pair of compasses (or as it is commonly called "a compasses") and an unmarked straightedge.

You will see when we discuss the history of π in the next chapter that it was in large part this quest for squaring the circle that resulted in more and more accurate approximations for the value of π. Despite Lindemann's work and that of others, many enthusiasts keep sending their "proofs" for squaring the circle to universities every year. They don't, or can't, accept the notion of the impossibility of squaring a circle. They cannot understand that when something has been proved to be impossible, it doesn't mean that we just weren't able to figure out how to do it; rather, we proved it is impossible to do.

Sharpening Our Intuition with π

Even in everyday life, knowledge of what π really represents can heighten our understanding of our faulty perceptions. Here is a simple illustration of how this knowledge lets us see the geometric world more objectively. Take a tall and narrow cylindrical drinking glass. Ask a friend if the circumference is greater or less than the height. The glass should be chosen so that it would "appear" to have a longer height than its circumference. (The typical tall narrow drinking glass fits this requirement.) Now ask your friend how she might test her conjecture (aside from using a piece of string). Recall for her that the formula for the circumference of a circle is $C = \pi d$ (π times the diameter). She should recall that $\pi \approx 3.14$ is the usual approximation, but we'll be even more crude and use $\pi = 3$. Thus the circumference will be 3 times the diameter, which can be easily "measured" with a stick or a pencil and then marked off 3 times along the height of the tall glass. Usually you will find that the circumference is longer than the height of the tall glass, even though it does not "appear" to be so. This little optical trick is useful to demonstrate the value of knowing the ratio of the circumference of a circle to its diameter, namely, π.

What the Bible Has as the Value of π

Let's stay with this "crude" approximation of π for a moment. You'll be surprised to know that for centuries scholars believed that this was the value that π was to have had in biblical times. For many years virtually all the books on the history of mathematics stated that in its earliest manifestation in history, namely, in the Old Testament of the Bible, the value of π is given as 3. Yet recent "detective work" shows otherwise.[12]

One always relishes the notion that a hidden code can reveal long-lost secrets. Such is the case with the common interpretation of the value of π in the Bible. There are two places in the Bible where the same sentence appears, identical in every way except for one word, which is spelled differently in the two citations. The description of a pool, or fountain, in King Solomon's temple is referred to in the passages that may be found in 1 Kings 7:23 and 2 Chronicles 4:2, and reads as follows:

And he made the molten sea[13] of ten cubits from brim to brim, round in compass, and the height thereof was five cubits; and *a line* of thirty cubits did compass it round about.

The circular structure described here is said to have a circumference of 30 cubits[14] and a diameter of 10 cubits. From this we notice that the Bible has $\pi = \frac{30}{10} = 3$. This is obviously a very primitive approximation of π. A late-eighteenth-century rabbi, Elijah of Vilna

---

12. Alfred S. Posamentier and Noam Gordon, "An Astounding Revelation on the History of π," *Mathematics Teacher* 77, no. 1 (January 1984): 52.

13. The "molten sea" was a gigantic bronze vessel for ritual ablutions in the court of the First Temple (966–955 BCE). It was supported on the backs of twelve bronze oxen (volume $\approx$ 45,000 liters).

14. A cubit is the distance from a person's fingertip to his elbow.

(1720–1797),[15] one of the great modern biblical scholars who earned the title "Gaon of Vilna" (meaning genius of Vilna), came up with a remarkable discovery, one that could make most history-of-mathematics books faulty if they say that the Bible approximated the value of π as 3. Elijah of Vilna noticed that the Hebrew word for "line measure" was written differently in each of the two biblical passages mentioned above.

In 1 Kings 7:23 it was written as קוה, whereas in 2 Chronicles 4:2 it was written as קו. Elijah applied the ancient biblical analysis technique (still used by talmudic scholars today) called gematria, where the Hebrew letters are given their appropriate numerical values according to their sequence in the Hebrew alphabet, to the two spellings of the word for "line measure" and found the following. The letter values are ק = 100, ו = 6, and ה = 5. Therefore, the spelling for "line measure" in 1 Kings 7:23 is קוה = 5 + 6 + 100 = 111, while in 2 Chronicles 4:2 the spelling קו = 6 + 100 = 106. Using gematria in an accepted way, he then took the ratio of these two values: $\frac{111}{106}$ = 1.0472 (rounded to four decimal places), which he considered the necessary "correction factor." By multiplying the Bible's apparent value of π, 3, by this "correction factor," one gets 3.1416, which is π correct to four decimal places! "Wow!" is a common reaction. Such accuracy is quite astonishing for ancient times. Moreover, remember how just getting π = 3.14 using string measurements was quite a feat. Now imagine getting π accurate to four decimal places. We would contend that this would be nearly impossible with typical string measurements. Try it if you need convincing.

Let's keep our focus on our effort to just getting acquainted with π. For the moment we are merely surveying the nature of π and what it means.

---

15. In those days Vilna was in Poland, while today the town is named Vilnius and is in Lithuania.

Where the Symbol π in Mathematics Came From

You may be wondering by now where mathematicians actually got the idea to represent the ratio of the circumference of a circle to its diameter with the Greek letter π. According to the well-known mathematics historian Florian Cajori (1859–1930), the symbol π was first used in mathematics by William Oughtred (1575–1660) in 1652 when he referred to the ratio of the circumference of a circle to its diameter as $\frac{\pi}{\delta}$, where π represented the periphery[16] of a circle and δ represented the diameter. In 1665 John Wallis used the Hebrew letter מ (mem), to equal one-quarter of the ratio of the circumference of a circle to its diameter (what, today, we would refer to as $\frac{\pi}{4}$).

In 1706 William Jones (1675–1749) published his book *Synopsis palmariorum matheseos*, in which he used π to represent the ratio of the circumference of a circle to its diameter. This is believed to have been the first time that π was used as it is defined today. Yet, Jones's book alone would not have made the use of the Greek letter π to represent this geometric ratio as popular as it has become today. It was the legendary Swiss mathematician Leonhard Euler, often considered the most prolific writer in the history of mathematics, who is largely responsible for today's common use of π. In 1736 Euler began using π to represent the ratio of the circumference of a circle to its diameter. But not until he used the symbol π in 1748 in his famous book *Introductio in analysin infinitorum* did the use of π to represent the ratio of the circumference of a circle to its diameter become widespread.

---

16. Note well, this is *not* what π was later on to represent.

Euler

Euler is not only the most prolific contributor to the development of mathematics, he also has given us quite a few symbols that are still commonly used today. These include the following:

$f(x)$, for the common notation for a mathematical function
e, for the base of natural logarithms
a, b, c, for the lengths of the sides of a triangle
s, for the semiperimeter of a triangle
r, for the length of the radius of the inscribed circle of a triangle
R, for the length of the radius of a circumscribed circle of a tri
    angle
Σ, for the summation sign
i, for the value of $\sqrt{-1}$

Euler discovered one of the most famous formulas in mathematics. It involves the symbols e, i, and π in the following way: $e^{i\pi} = -1$. The mathematicians Edward Kasner and James Newman, in their book *Mathematics and the Imagination*, make the following statement about this formula: "Elegant, concise, and full of meaning, we can only reproduce it and not stop to inquire into its implications. It appeals equally to the mystic, the scientist, the philosopher, and the mathematician. For each it has its own meaning."[17] They go on to tell the anecdote about the nineteenth-century Harvard mathematician Benjamin Peirce, who having come upon the formula "turned to his students and made a remark which supplies in dramatic quality and appreciation what it may lack in learning and sophistication: 'Gentlemen,' he said, 'that is surely true, it is absolutely paradoxical; we cannot understand it,

---

17. Edward Kasner and James Newman, *Mathematics and the Imagination* (New York: Simon and Schuster, 1940), p. 103.

and we don't know what it means, but we have proved it, and therefore, we know it must be the truth.'" So it is with much of mathematics—we prove something and it becomes accepted—understanding can follow!

Since Euler is the father of the symbol that has the title role of this book, we ought to take a glimpse into his interesting life history. Born in Basel, Switzerland, in 1707, he was initially taught mathematics by his father, who himself studied under the famous mathematician Jakob Bernoulli. This connection served him well, for as the father noticed his son's proclivity for the subject, he arranged for him to study with Jakob Bernoulli's son (also a famous mathematician) Johann Bernoulli. Through the influence of the Bernoulli family, Euler got a position at age twenty with the Russian Academy in St. Petersburg, where he stayed for fourteen years. During this time he rose to the position of chief mathematician. Although Euler spent the next twenty-five years at the Prussian Academy, he never lost touch with the Russian Academy, to which he returned for the remaining seventeen years of his life.

It is well known that Ludwig van Beethoven spent the last years of his life totally deaf and, despite this enormous handicap, continued to produce magnificent musical compositions—most notably his Ninth Symphony. An analogous calamity struck Euler. Clearly the requirement of being able to see is essential to do mathematics, as one's ability to hear sound is imperative to being able to compose music. Euler lost the sight in his right eye as early as 1735, yet he was unimpaired in his mathematical output. This, by the way, accounts for the poses that we see in pictures of Euler (see fig. 4).

Soon after his return to St. Petersburg at the invitation of Catherine the Great, Euler became blind, yet, largely due to his incredible memory, remained just as productive. However, now he had to dictate his ideas to his secretary. Euler's record-setting output is about 530 books and articles during his lifetime, and many more manuscripts were left to posterity. These continued to appear in the *Proceedings of the St. Petersburg Academy* for forty-seven years after his death. It is estimated that his total production was about 886 books and articles.[18] Truly astonishing—especially since he himself could not see many of these!

Leonhard Euler
Fig. 1-4 a

---

18. Howard Eves, *An Introduction into the History of Mathematics*, 5th ed. (New York: CBS College Publishing, 1983).

Leonhard Euler
Fig. 1-4 b-c-d-e

A π Paradox

We mentioned earlier that the interest taken in π is partially due to its ubiquity. It quickly transcends the ratio that is used to define it. The concept of π pops up in places where we are left truly perplexed. One such involves an entertaining illustration of a paradox in geometry. This example may also be considered a geometric fallacy. Follow along as we explain it, and see if you can determine "what's wrong here."

In the figure below, the smaller semicircles extend from one end of the large semicircle's diameter to the other.

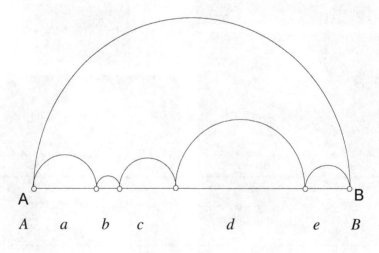

Fig. 1-5

Let us begin by showing that the sum of the arc lengths of the smaller semicircles is equal to the arc length of the larger semicircle.

That is, the sum of the smaller semicircles equals

$$\frac{\pi a}{2}+\frac{\pi b}{2}+\frac{\pi c}{2}+\frac{\pi d}{2}+\frac{\pi e}{2}=\frac{\pi}{2}(a+b+c+d+e)=\frac{\pi}{2}(AB),$$

which is the arc length of the large semicircle, since the large semi-circle's arc length is one-half the diameter (AB) times π. This may not "appear" to be true, but it is! Let's imagine that we were to increase the number of smaller semicircles along the fixed line segment AB.

This sort of progression of increasing the number of semicircles can be seen in the following figures.

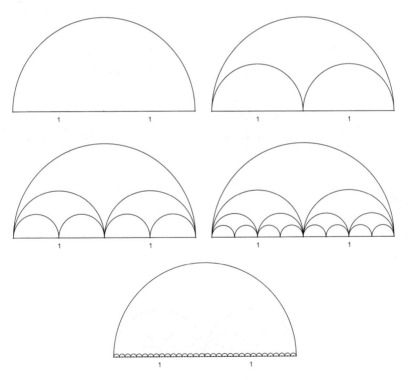

Fig. 1-6 a-b-c-d-e

They, of course, get smaller. The sum of these smaller semicircular arcs "appears" to be approaching the length of the diameter *AB* (referring back to the earlier figure), but, in fact, does not! Suppose the diameter of the large semicircle is 2; then the semicircular arc length is π. If the sum of the increasingly smaller semicircles becomes π, then π is equal to 2, the length of the diameter. Impossible! (By now we know that even in the Bible it was recognized that π was at least 3.) So what "appears" to be true from the diagram and some "logical" extension of it, namely, that the semicircular arc length is equal to the straight-line segment, leads to an absurd conclusion. It does *not* follow, however, that the *sum* of the semicircles approaches the *length* of the limit, which in this case is *AB*. This "apparent limit sum" is absurd, since the shortest distance between points *A* and *B* is the length of segment *AB*, not the semicircle arc *AB* (which equals the sum of the smaller semicircles). Just as this faulty reasoning led us to a weird conclusion, some faulty thinking led some Indiana legislators to hold a place in the history of mathematics with some rather strange actions. Read on.

Legislating π

The value of π has vexed mathematicians and others for centuries, yet perhaps the most outrageous attempt to "nail down" the value of π occurred in Indiana in 1897. A physician there by the name of Edward Johnson Goodwin (1828–1902) wrote a paper on measurements of the circle and convinced his local legislative representative, Taylor I. Record, to introduce it as a bill in the legislature. The epoch-making suggestion that he put to Taylor I. Record was this: If the state would pass an act recognizing his, Goodwin's, discovery, then he would allow all Indiana textbooks to use it without paying him a royalty.

He had already copyrighted his findings in various European countries and in the United States. His attempt to present his findings at the Columbian Exposition in Chicago in 1893 failed however. He did publish a monograph in the *American Mathematical Monthly*, a new journal, eager to accept almost anything in its first year. From Goodwin's monograph one can get as many as nine different values of π. These were calculated by mathematician David Singmaster[19] to be:

π = 4, 3.160494, 3.232488, 3.265306, 3.2, 3.333333, 3.265986, 2.56, and 3.555556.

On January 18, 1897, the monograph was entered into the legislature as House bill no. 246.

A bill for an act introducing a new mathematical truth and offered as a contribution to education to be used only by the State of Indiana free of cost by paying any royalties whatever on the same, provided it is accepted by the official action of the legislature of 1897.

At first it was accepted without negative vote in the House of Representatives of Indiana. It could have attained legal status, where all other states would have to pay for the right to this "exact value" of π. Till then, clearly, one needed to pay nothing for mathematical truths.

By legislating the value of π, Goodwin believed he would put the problem of determining the value of π to rest. Fortunately, through the newspapers in Indianapolis, Chicago, and New York, much ridicule was cast upon this silly bill, and the Indiana Senate eventually killed it. This is just one of many unreasonable efforts to secure a value for π.

---

19. David Singmaster, "The Legal Values of Pi," *Mathematical Intelligencer* (New York: Springer Verlag) 7, no. 2 (1985): 69–72.

π in Probability

π shows up in some of the strangest places. To whet your appetite, we offer one example of how π seems, amazingly enough, to invade fields of mathematics that apparently have nothing to do with geometry, such as probability.

The French naturalist Georges Louis Leclerc, Comte de Buffon (1707–1788) is primarily remembered for his work to popularize the natural sciences in France, and his *Histoire naturelle* (1749–1767) is still prized today, largely because of the exceptional beauty of the illustrations. In it all the known facts of the natural sciences are eloquently discussed, and Buffon even foreshadowed the theory of evolution. Yet in mathematics he is remembered for two things: his French translation of Newton's *Method of Fluxions*, the forerunner of today's calculus, and more so even for the "Buffon needle problem."[20] It is the latter that is of particular interest to us here.

In his "Essai d'arithmétique morale," published in 1777, he proposes a very intriguing phenomenon relating π to probability. It goes this way: suppose you have a piece of paper with ruled parallel lines throughout, equally spaced (at a distance d between lines), and a thin needle of length l (where $l < d$). You then toss the needle onto the paper many times. Buffon claimed that the probability that the needle will touch one of the ruled lines is $\frac{2l}{\pi d}$. Since Buffon was a man of wealth and had much time to spare, he tried this experiment with thousands of tosses to substantiate his conclusions. For the next thirty-five years this problem was essentially forgotten until the preeminent mathematician Pierre Simon Laplace (1749–1827) popularized it. We must bear in mind that Laplace was one of the greatest French mathematicians, and in 1812 he pub-

---

20. For a more complete discussion of Buffon's needle problem see Lee L. Schroeder, "Buffon's Needle Problem: An Exciting Application of Many Mathematical Concepts," *Mathematics Teacher* 67, no. 2 (1974): 183–86.

lished a major work in probability, *Théorie analytique des probabilities*, which gave him much prominence in the field.

You may want to try Buffon's experiment yourself. Begin by simplifying the problem (without any loss of generality) by letting $l = d$, so that the probability of the needle (now with a length equal to the space between the lines) touching one of the lines is $\frac{2}{\pi}$. That is, $\pi = \frac{2}{P}$, where P is the probability that the needle will intersect the line, which is

$$P = \frac{number\ of\ line\text{-}touching\ tosses}{number\ of\ all\ tosses}$$

So to calculate π this way, just toss the needle and tally the line-touching tosses and the total number of tosses. Then put them into this formula:

$$\pi = \frac{2 \times number\ of\ all\ tosses}{number\ of\ intersection\ tosses}$$

The more tosses you have, the more accurate your estimate of π should be. In 1901 the Italian mathematician Mario Lazzarini tried this with 3,408 tosses of the needle and got $\pi = 3.1415929$, an amazing accuracy. You might also try to have a computer simulate the needle tossing. It's much easier that way. In any case, this is by far not the most accurate way to calculate the value of π. It is, however, quite novel. Just think about it. The probability of a tossed needle intersecting a line is related to π, the ratio of the circumference of a circle to its diameter.

We will next provide you with a simple tour through the long journey mathematicians have taken over four thousand years to get an increasingly more accurate estimate for the value of π. This history of π will take some large leaps; however, we will highlight the more significant and easily understood methods developed over the millennia.

Chapter 2

The History of π

In the Beginning

The story of π probably goes much further back in time than we can document through written records. Somewhere in the past, after a wheel (or any truly circular object) was invented, the circumference was probably measured for the sake of comparison. Perhaps in the early days it was important to measure how far a wheel would travel in one revolution. This might have been done by rolling the wheel on the ground and marking off the distance it rolled in exactly one revolution (without slippage, of course) or with something resembling a string placed along it. The diameter, a much easier dimension to measure, since it merely required placing a straight stick or rule alongside it and marking off its length, was probably also noted. We can assume that these two measurements were compared for various circular objects. This was likely the beginning of the establishment of comparison between the two measurements that seem related to each other. Was there some sort of common difference or common ratio between their lengths? Each time this comparison showed that the circumference was just a bit more than three times as long as the diameter. The question that perplexed individuals over the millennia was how much more than three times the diameter was the circumference? That would indicate that the relationship was one of a ratio. The history of π is the quest to find the ratio between the circumference of a circle and its diameter.

The Ancient Egyptians

Frequent measurements probably showed that the part exceeding three times the diameter appeared to be about one-ninth of the diameter. We can assume this from the famous Rhind Papyrus, written by Ahmes, an Egyptian scribe, about 1650 BCE.[1] He said that if we construct a square with a side whose length is eight-ninths of the diameter of the circle, then the square's area will be equal to that of the circle. At this point, you can see there was no reason to find the ratio of the circumference to the diameter. Rather, the issue was to construct a square, using the classical tools (an unmarked straightedge and a pair of compasses), with the same area as that of a given circle. This became one of the three famous problems of antiquity.[2] Although we know today

---

1. This was a mathematical practical handbook, containing eighty-five problems copied by the scribe Ahmes from previous works. Alexander Henry Rhind, a Scottish Egyptologist, purchased this eighteen-foot-long (one-foot-wide) manuscript in 1858, which is now in the collection of the British Museum. This is one of our primary sources of information about the Egyptian mathematics of the times.

2. The other two famous problems of antiquity are using only an unmarked straightedge and a pair of compasses to construct a cube with twice the volume of a given cube and using these same tools to trisect any angle.

that this is an impossible construction,[3] it, nonetheless, fascinated mathematicians for centuries. It was the effort to construct a square with an area equal to that of a given circle that produced the early approximations of π. For example, if we inspect the process used in the Rhind Papyrus, we can deduce how close the ancient Egyptians were to the true value of π. We will now try to replicate their work.

We will begin with a circle with diameter d. According to the above stipulations, the side of the square would then be $\frac{8}{9}d$.

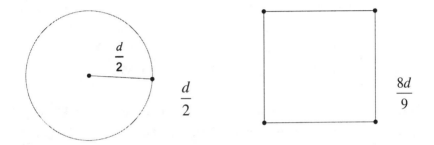

Fig. 2-1

We know from today's knowledge about circles that the area of the circle[4] is πr^2, which for this circle gives us[5]

$$\pi\left(\frac{d}{2}\right)^2 = \pi\frac{d^2}{4}$$

---

3. As noted earlier, the impossibility of constructing a square with area equal to that of a given circle was conjectured for many years, but was first proved conclusively in 1882 by the German mathematician Carl Louis Ferdinand Lindemann (1852–1939).

4. We mentioned that the symbol π was not used to represent the ratio of the circumference of a circle to the diameter until more than three thousand years later. However, for convenience and to avoid confusion, we will use the symbol π already at this early stage.

5. The equal sign (=) was first used by the English physician and mathematician Robert Recorde (1510?–1558) in "The Whetstone of Witte" (1557), when he said that "noe .2. thynges, can be moare equalle" than the two parallel lines that make up the equal sign.

The area of the square is simply

$$\left(\frac{8d}{9}\right)^2 = \frac{64d^2}{81}$$

Since Ahmes assumed these to be equal, we get the following equation:

$$\pi\frac{d^2}{4} = \frac{64d^2}{81}$$

$$\frac{\pi}{4} = \frac{64}{81}$$

So

$$\pi = \frac{256}{81} = 3.1604938271\overline{604938271604938271}$$

This is a reasonably close approximation of what we know the value of π to be by using our modern methods.

Just Before the Common Era

We now take a big leap in time to the Babylonians, which spans from 2000 BCE to about 600 BCE. In 1936 some mathematical tablets were unearthed at Susa (not far from Babylon).[6] One of these compares the perimeter of a regular hexagon[7] to the circumference of its circumscribed circle. The way they did this led today's mathematicians to deduce that the Babylonians used $3\frac{1}{8}$ = 3.125 as their approximation for π. How does this compare to the Egyptians' approximation for π? It is just a very little bit closer.

6. Today, easiest located as the region between the Tigris and Euphrates rivers.

7. A regular polygon (in this case a hexagon, a polygon of six sides) is one where all the sides are the same length and all the angles are equal.

As we progress through the early history of the development of the ratio (π) of the circumference to the diameter of a circle, we come upon the Bible (Old Testament) written about 550 BCE, where the Talmud's books of Kings and Chronicles describe King Solomon's water basin (or well) and give us the impression that they believed π = 3. However, we discussed earlier (see pages 27–28) the notion that there might have been a hidden value in these writings yielding the value π = 3.1416, even a more accurate value than the earlier ones.

One of the biggest challenges facing these ancient mathematicians was to be able to measure a circular figure (even parts of circle) in terms of straight lines. This was essentially the problem to be solved in "squaring the circle," that is, constructing a side of a square whose area is equal to that of a given circle. Circular arcs and straight lines could not find a common measure. There was always "something left over" when trying to compare these two types of measurement. Hippocrates of Chios, another Greek mathematician who flourished about 430 BCE, was the first to be able to show that areas of lunes (i.e., areas bounded by circular arcs) can be equal to the area of a rectilinear figure, such as a triangle.[8] Although Hippocrates' works are lost, we shall show an example that may have been similar to his. In other words, we will show an example where a region bounded by circular arcs can be exactly equal to a region bounded by straight lines.

To tackle this, let's first recall the famous Pythagorean theorem. It states that *the sum of the squares of the legs of a right triangle is equal to the square of the hypotenuse.* This can be stated a bit differently with the same effect: *The sum of the squares* on *the legs of a right triangle is equal to the square* on *the hypotenuse.* Geometrically this can

---

8. A rectilinear figure is one bounded by straight line segments.

be seen in figure 2-2, where the sum of the areas of the two shaded squares is the same as the larger area of the unshaded square.

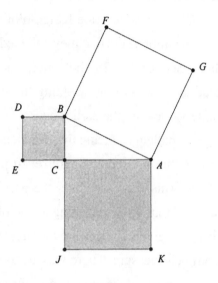

Fig. 2-2

This can then be restated as *the sum of the areas of the squares on the legs of a right triangle is equal to the area of the square on the hypotenuse,* which then draws us a big step forward to a generalization that will allow us to replace the squares with any similar polygons, as long as they are placed in corresponding orientation. That is, the corresponding sides of these similar polygons must coincide with the sides of the right triangle on which they are placed. We can then make the following generalization:

The sum of *the areas of* the *similar polygons* on the legs of a right triangle is equal to *the area of* the *similar polygon* on the hypotenuse.

For our purposes, we will use semicircles to represent our similar polygons, since all semicircles are the same shape, and hence, similar. This will then read as follows:

The sum of *the areas of* the semicircles on the legs of a right triangle is equal to *the area of* the semicircle on the hypotenuse.

This extension of the Pythagorean theorem can be proved, by considering the three sides of the right triangle to be $2a$, $2b$, and $2c$. Then the areas of the three semicircles are $\frac{\pi a^2}{2}$, $\frac{\pi b^2}{2}$, and $\frac{\pi c^2}{2}$. Let's see if this relationship holds. That is, is $\frac{\pi a^2}{2} + \frac{\pi b^2}{2} = \frac{\pi c^2}{2}$? Dividing through by the common factor $\frac{\pi}{2}$ gives us $a^2 + b^2 = c^2$, which we know will result by applying the Pythagorean theorem to this right triangle. That is, we get $4a^2 + 4b^2 = 4c^2$, which is then $a^2 + b^2 = c^2$. Thus, for the figure below (fig. 2-3), we can say that the areas of the semicircles relate as follows:

Area P = Area Q + Area R

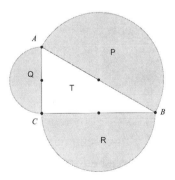

Fig. 2-3

Suppose we now flip semicircle P over the rest of the figure (using $\overline{AB}$ as its axis). We would get a figure as shown below. Notice that the flipped-over semicircle now forms four new regions marked L_1, L_2, J_1, and J_2.

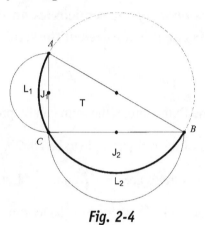

Fig. 2-4

Let us now focus on the lunes formed by the two semicircles. We mark them L_1 and L_2.

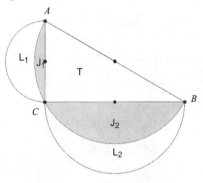

Fig. 2-5

When we extended the Pythagorean theorem (above) to semicircles instead of squares, we established that

$$Area\ P = Area\ Q + Area\ R$$

In the figure above, keeping in mind the largest semicircle's new position—that being flipped over the triangle—that same relationship can be written as follows:

Area J₁ + Area J₂ + Area T = Area L₁ + Area J₁ + Area L₂ + Area J₂

Take a moment to convince yourself of this relationship.

If we subtract *Area J₁ + Area J₂* from both sides, we get the astonishing result:

$$Area\ T = Area\ L_1 + Area\ L_2$$

That is, we have the area of a rectilinear[9] figure (the triangle) equal to the sum of the areas of some nonrectilinear figures (the lunes).[10] This is a very profound result, since it is at the crux of one of the most vexing issues in mathematics—that of finding equality between measurements of circles and rectilinear figures. As we said before, this was one of the challenges that faced ancient mathematicians as they tried to square the circle.

There is a nice three-dimensional example in which a sphere has the same volume as a rectilinear figure, namely, a tetrahedron, which is a solid figure with four faces (planes). So as not to disturb the continuity in this chapter, we provide this discussion in appendix A. (See page 293.)

Although the circle-ratio π is indispensable in the calculation of the area of circles (or semicircles), the famous Pythagorean theorem eliminates π from the comparison of areas of semicircles on the three sides of a right triangle.

---

9. A rectilinear figure is one that is only bounded by straight lines.
10. A lune in the plane is a closed figure bordered by circular arcs.

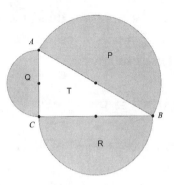

Fig. 2-6

Let us return to the relationship we established earlier, namely, that

$$Area\ P = Area\ Q + Area\ R$$

so that we get

$$\frac{1}{2} \bullet \pi \bullet \left(\frac{c}{2}\right)^2 = \frac{1}{2} \bullet \pi \bullet \left(\frac{b}{2}\right)^2 + \frac{1}{2} \bullet \pi \bullet \left(\frac{a}{2}\right)^2$$

where $a = BC$, $b = AC$, and $c = AB$

This gives us

$$\frac{\pi}{8} \bullet c^2 = \frac{\pi}{8} \bullet b^2 + \frac{\pi}{8} \bullet a^2,$$

which reduces to $c^2 = b^2 + a^2$. Notice the π disappeared![11]

Euclid's *Elements* (ca. 300 BCE), clearly the first and most comprehensive geometry book ever written, also made a contribution to the history of π. In Book XII, Proposition 2, Euclid states and proves that "circles are to each other as the squares on the

---

11. We simply multiplied each term by $\frac{8}{\pi}$.

diameters." This was probably taken from Hippocrates (not to be confused with the physician, Hippocrates of Cos). This is particularly significant because for the first time it establishes that there is, in fact, a constant, such as π, that relates the circumference to the diameter of a circle. What is being said here might be clearer if shown symbolically:

$$\frac{\text{area of circle 1}}{\text{area of circle 2}} = \frac{(\text{diameter of circle 1})^2}{(\text{diameter of circle 2})^2}$$

A simple (and legitimate) algebraic manipulation lets us change the proportion above to read as

$$\frac{\text{area of circle 1}}{(\text{diameter of circle 1})^2} = \frac{\text{area of circle 2}}{(\text{diameter of circle 2})^2} = \text{some constant value}$$

Let's take just one of these fractions and set it equal to the constant, which today we know[12] is actually $\frac{\pi}{4}$.

Another way of writing this is that the area of circle 1 equals (diameter of circle 1)^2 × (some constant value)

$$= d^2 \frac{\pi}{4} = (2r)^2 \frac{\pi}{4} = 4r^2 \frac{\pi}{4} = \pi r^2$$

This says that the area of a circle is equal to some constant, say $\frac{\pi}{4}$, times the square of the diameter (or for that matter twice the radius). Eventually, it leads us to the formula for the area of a circle. Actually this work of Euclid only hints at the possible awareness of a constant π. We followed it to (what we know today as) the correct representation of π.

---

12. Using our modern knowledge, we can represent this as $\frac{\pi r^2}{(2r)^2} = \frac{\pi r^2}{4r^2} = \frac{\pi}{4}$.

Archimedes' Contributions

One of the greatest contributors in the early history of mathematics was Archimedes, born in Syracuse (Sicily) about 287 BCE, the son of the astronomer Phidias. For a time he studied with the successors of Euclid in Alexandria, Egypt. There he also met Conon of Samos, for whom he had high regard as an astronomer and mathematician, and Eratosthenes of Cyrene, with whom he corresponded for years after leaving Egypt. His contributions to mathematics and physics are legendary. We will focus only on one small part of his work: that involving the circle and π.

Not until Archimedes was there a rigorous connection between the circumference of a circle and its area. This can be found in Archimedes' *Measurement of the Circle*. In this important book there are three propositions regarding the circle that have had a role in the historical development of the value of π. We shall present these three propositions along with a bit of explanation of each.

1. The area of a circle is equal to that of a right triangle where the legs of the right triangle are respectively equal to the radius and circumference of the circle.

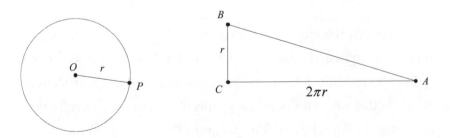

Fig. 2-7

The area of the circle is the familiar πr^2, and the area of the right triangle (which is one-half the product of its two legs) is

$$\frac{1}{2}(r)(2\pi r) = \pi r^2$$

Although Archimedes stated this in a somewhat convoluted way, it is amazing that he hit the formula that we accept today right on the head!

2. The ratio of the area of a circle to that of a square with side equal to the circle's diameter is close to 11:14.

To investigate this proposition, we will set up the ratio as it is given to us.

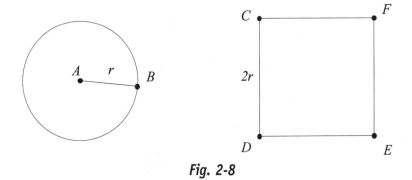

Fig. 2-8

The area of the circle is πr^2, and the area of the square (whose side is 2r) is $(2r)^2 = 4r^2$. The ratio of these is

$$\frac{\pi r^2}{4r^2} = \frac{\pi}{4} = \frac{11}{14},$$

as was stated in the proposition. When we simplify this proportion we get

$$\pi = \frac{44}{14} = \frac{22}{7},$$

which should remind you of another very familiar approximation of π.

3. The circumference of a circle is less than $3\frac{1}{7}$ times its diameter but more than $3\frac{10}{71}$ times the diameter.

Let us take a quick look at how Archimedes actually came to this conclusion. (A more-detailed discussion of his work will be found in chapter 3.) What Archimedes did was to inscribe a regular hexagon[13] in a given circle and circumscribe a regular hexagon about this same circle. He was able to find the areas of the two hexagons and then knew that the area of the circle had to be somewhere between these two areas.

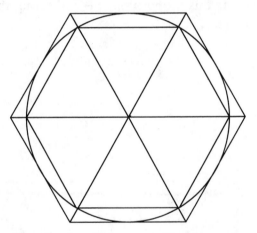

Inscribed and circumscribed hexagons

Fig. 2-9

He then repeated this with regular dodecagons (twelve-sided regular polygons) and again calculated the area of each, realizing that the circle's area had to be between these values, and more closely "sandwiched in," to use a modern analogy.

---

13. A regular hexagon is a six-sided polygon that has all sides and angles equal.

This was then done for twenty-four-sided regular polygons, forty-eight-sided regular polygons, and ninety-six-sided regular polygons, each time getting closer and closer to the area of the circle. Mind you, this was done before the Hindu number system was used in the Western world—no mean feat of calculations! Archimedes finally concluded that the value of π is larger than $3\frac{10}{71}$ and less than $3\frac{1}{7}$. How does this compare to our known value of π? We change these fractions to decimal form so that we can make a comparison of their values to what we know today as the true value of π.

Therefore, since

$$3\tfrac{10}{71} = 3.\overline{140845070422535211267605633802816}90$$

and

$$3\tfrac{1}{7} = 3.142857\overline{142857142857}$$

we can see how well Archimedes placed the value of π:

$$3.\overline{140845070422535211267605633802816}90 < \pi < 3.\overline{142857}$$

This is consistent with what we know as the value of π today, 3.14159265358979323846264338327950288419716939937510 58 ...(taken to over fifty decimal places).

Our known value of π is nicely squeezed in between the two values that Archimedes used as boundaries.

For now, we can leave this with the notion that he saw a circle as the limit of the ever-increasing number of sides of a regular polygon of a fixed perimeter.

Two closer approximation values have been found, according to Heron of Alexandria (75–110), in a lost document of Archimedes:

$$\frac{211{,}872}{67{,}441} < \pi < \frac{195{,}882}{62{,}351},$$

which places π in the interval **3.14 15 90** . . . **< π < 3.14 16 01** . . .

In the passing years, the approximations became ever closer to the value of π, so that in 200 BCE Apollonius of Perge (262–190 BCE), a competitor of the great Archimedes, seemed to have discovered an even better approximation for π than that of Archimedes:

$$\pi \approx 3\frac{177}{1{,}250} = \frac{3{,}927}{1{,}250} = 3.1416$$

Regardless, we still consider Archimedes to be one of the major contributors to the history of mathematics.

Archimedes' life proceeded quietly up to his death in 212 BCE. He was killed defending his hometown of Syracuse during the Second Punic war. Archimedes was believed to have said to a Roman soldier, who came to summon him to the emperor Marcellus, and whose shadow covered one of his drawings in the sand: "Don't disturb my circles" ("Noli turbare circulos meos"), whereupon the soldier stabbed him to death. Archimedes requested that his tombstone be decorated with a sphere contained in the smallest possible cylinder and inscribed with the ratio of the sphere's volume to that of the cylinder.[14] Archimedes had considered the discovery of this ratio the greatest of all his accomplishments.

---

14. This may be found in Archimedes' book *On the Sphere and the Cylinder*.

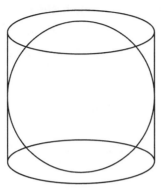

Fig. 2-10

The relationship between these two solids is truly unusual. The ratio of their volumes and the ratio of their surface areas is the same! Both are 2:3. We can easily calculate these with our current knowledge about the formulas for these various figures.

The formula for the volume of a sphere is $\frac{4}{3}\pi r^3$.[15] The volume of the cylinder is obtained by taking the area of the base and multiplying it by the height:

$$\left(\pi r^2\right)\left(2r\right) = 2\pi r^3 = \frac{6}{3}\pi r^3$$

(we wrote 2 as the fraction $\frac{6}{3}$ to make the comparison easier). Thus the ratio of the volumes of the sphere to the cylinder is

$$\frac{\frac{4}{3}\pi r^3}{\frac{6}{3}\pi r^3} = \frac{2}{3}$$

---

15. This formula was first published by Archimedes in his book *On the Sphere and the Cylinder*.

Now let's compare the surface areas of the two solids. The formula for the surface area of a sphere is $4\pi r^2$. The surface area of the cylinder is found by adding the areas of the two bases to the lateral area of the cylinder:

$$(2)\left(\pi r^2\right) + (2r)(2\pi r) = 6\pi r^2$$

Comparing these two surface areas, we get

$$\frac{4\pi r^2}{6\pi r^2} = \frac{2}{3}$$

Lo and behold, the same ratio—truly amazing!

In his book *On the Sphere and the Cylinder*, Archimedes also stated that "a sphere is four times as great as a cone with a great circle of the sphere as its base and with its height equal to the radius of the sphere."[16] This can be extended by the comparison of the cone to the cylinder that contains the sphere. We can easily establish Archimedes' proposition above, for the cone with base radius r and height r has a volume equal to

$$\frac{1}{3}\left(\pi r^2\right)(r) = \frac{1}{3}\pi r^3,$$

which is $\frac{1}{4}$ of the volume of the sphere of radius r.

Now if we double the length of the height of this cone so that it can be inscribed in the cylinder of equal height, then its volume will be

$$\frac{1}{3}\left(\pi r^2\right)(2r) = \frac{2}{3}\pi r^3,$$

or one-half the volume of the sphere.

---

16. The great circle of a sphere is the largest circle that can be drawn on a sphere—or to put it more simply, if we were to cut a sphere into two hemispheres, their base would be a great circle of the sphere.

Thus we can represent that the volumes of

cone $\left[\dfrac{2}{3}\pi \cdot r^3\right]$,

sphere $\left[\dfrac{4}{3}\pi \cdot r^3\right]$,

and cylinder $\left[2\pi \cdot r^3\right]$

with the same base are in the ratio of 1:2:3.

Archimedes is still revered today, hailed as the greatest thinker of his time, with countless ingenious inventions and mathematical achievements. As evidence of his popularity, on October 29, 1998, a book of his, on the calculation of areas and volumes, brought $2 million at a Christie's auction.

Although we assumed earlier that in ancient times circumferences might have been measured by the distance a wheel traveled in one revolution, Marcus Vitruvius Pollio, more commonly known today as Vitruvius, a Roman architect and engineer, used this method to calculate π as $3\frac{1}{8}$ = 3.125. This was not exactly a step forward, given that he wrote his book *da Architectura* in the year 20 BCE.

The Beginning of the Common Era

We now get a bit closer to the true value of π with the great astronomer, geographer, and mathematician Claudius Ptolemaeus, popularly known as Ptolemy (ca. 83 CE–ca. 161 CE), who about

150 CE wrote an astronomical treatise, *Almagest*. He used the sexagesimal system[17] to get

$$\pi = 3 + \frac{8}{60} + \frac{30}{60^2} = 3\frac{17}{120} = 3.141666... = 3.141\overline{6} \approx 3.14167$$

This is the most accurate result after Archimedes.

The issue of establishing the irrationality of π was not settled until the eighteenth century (as we will see a bit later). However, it was anticipated by the great Jewish philosopher Maimonides (1135–1204)[18] in his commentary on the Bible, which states:

> You need to know that the ratio of the circle's diameter to its circumference is not known and it is never possible to express it precisely. This is not due to a lack in our knowledge, as the sect called Gahaliya [the ignorants] thinks; but it is in its nature that it is unknown, and there is no way [to know it], but it is known approximately. The geometers have already written essays about this, that is, to know the ratio of the diameter to the circumference approximately, and the proofs for this. This approximation, which is accepted by the educated people, is the ratio of one to three and one seventh. Every circle, whose diameter is one handbreadth, has in its circumference three and one seventh handbreadths, approximately. As it will never be perceived but approximately, they [the Hebrew sages] took the nearest integer and said that every circle whose circumference is three fists is one fist wide, and they contented themselves with this for their needs in the religious law.[19]

---

17. A number system using a base of 60, instead of the decimal system that uses the base 10.
18. His actual name was Moses ben Maimon, and he wrote commentaries on the Bible as well as treatises on logic, mathematics, medicine, law, and theology. He became rabbi of Cairo in 1177.
19. *Mishna* (Mishna Eruvin I 5), Mo'ed section (Jerusalem: Me'orot, 1973), pp. 106–107.

The Chinese Contributions

Meanwhile, in China, independent investigations in geometry paralleled some of the work in the Western world. Liu Hui in 263 CE also used regular polygons with increasing numbers of sides to approximate the circle. However, he used only inscribed circles, while Archimedes used both inscribed and circumscribed circles. Liu's approximation of π was

$$\frac{3,927}{1,250} = 3.1416$$

and might have been more accurate than Archimedes' approximation since he used a decimal number system with a place value system. Also noteworthy about Liu's work is that he assumed the area of a circle is half the circumference times half the diameter. Let us take a closer look at this assumption. What Liu had assumed can be written symbolically as

$$\frac{1}{2}C \cdot \frac{1}{2}d = \frac{1}{2}(2\pi r) \cdot \frac{1}{2}(2r) = \pi r^2$$

Recognize this? Yes, this is the familiar formula for the area of a circle.

Yet perhaps the most accurate approximation of π for the next thousand years was that of the Chinese astronomer and mathematician Zu Chongzhi (429–500), who through various mysterious ways[20] came up with

$$\pi = \frac{355}{113} = 3.14159292035398230088495557522123893805$$

$$3097345132743362831858407079646017699115044 2477$$

$$8761061946902654867256637168141592920353982 3008$$

$$8495557522123893805309734513274336283185840 70796$$

$$4601769911504424778761061946902654867256637 168$$

which continues by repeating every 112 places.

The Beginning of the Renaissance

Our next stop in tracing the history of π must be with Leonardo Pisano (1170–1250), better known as Fibonacci. Though a citizen in the city-state of Pisa, he traveled extensively throughout the Middle East and brought back to Italy a new understanding of and procedure in mathematics. In his famous book, *Liber abaci*, first published in 1202, he introduced the Hindu number system that we use today. It was the first published mention of this system in western Europe. It also contains the famous rabbit problem that produced the well-known Fibonacci numbers.[21] In 1223 he wrote *Practica geometriae*, where, making use of a regular polygon of ninety-six sides, he computed the value of π to be

$$\frac{1,440}{458\frac{1}{3}} = 3.141818181818181818181818181818,$$

which he obtained by taking the average between

$$\frac{1,440}{458\frac{1}{5}} = 3.14273243125272800663465735486687...$$

and

$$\frac{1,440}{458\frac{4}{9}} = 3.14105671352399418322830828899...$$

Although for his time his approximation was not as close as others, Fibonacci's contributions to the mathematics development of western Europe are legendary, especially for the times following the Dark Ages.

---

21. The Fibonacci numbers are 1, 1, 2, 3, 5, 8, 13, 21, 34, 55, 89,…, where each number after the first two is the sum of the two previous numbers.

The Sixteenth Century

Throughout the centuries many attempts at approximating the value of π continued, though the accuracy wavered back and forth. For example, at the turn of the sixteenth century the famous German artist and mathematician Albrecht Dürer (1471–1528) used an approximation for π of $3\frac{1}{8} = 3.125$, far less accurate than other approximations before that time.

A big change in the computation of π came in 1579, when the French mathematician François Viète (1540–1603), using the method developed by the Greeks, considered a regular polygon of $6 \cdot 2^{16} = 393,216$ sides and calculated π correct to nine decimal places. He also discovered the first use of an infinite product,[22] to determine the value of π.

$$\frac{2}{\pi} = \sqrt{\frac{1}{2}} \cdot \sqrt{\frac{1}{2} + \frac{1}{2}\sqrt{\frac{1}{2}}} \cdot \sqrt{\frac{1}{2} + \frac{1}{2}\sqrt{\frac{1}{2} + \frac{1}{2}\sqrt{\frac{1}{2}}}} \cdots$$

Viète calculated the value of π to be between: 3.1415926535 and 3.1415926537. Again, a new milestone in the long history of π was reached.

The process of letting regular polygons with enormous numbers of sides approach the dimensions of a circle continued. The next step forward in the quest to getting more accurate values of π came in 1593, when the Antwerp physician and mathematician Adriaen van Roomen,[23] using a regular polygon of 2^{30} sides (a polygon of 1,073,741,824 sides), calculated π to seventeen decimal places (of which the first fifteen decimal places were correct).

---

22. This refers to a product of an infinite number of terms following a given pattern.
23. Sometimes referred to by his Latin name, Adrianus Romanus.

The Seventeenth Century

The German mathematician Ludolph van Ceulen (1540–1610), who was intent on finding the true value of π, found its value accurate to twenty decimal places in 1596. His result was calculated from the perimeters of inscribed and circumscribed regular polygons of $60 \cdot 2^{33} = 515{,}396{,}075{,}520$ sides.

To achieve this, he had to discover some new theorems to carry out the calculations. The big step forward in this pursuit for a true value of π came in 1610, when Ludolph van Ceulen found the value of π to thirty-five decimal places, using a polygon of $2^{62} = 4{,}611{,}686{,}018{,}427{,}387{,}904$ sides. He was so devoted to (or we might say obsessed with) calculating the value of π and he made such great strides in that endeavor that, in his honor, π is sometimes referred to as the *Ludolphian number*. In addition, upon his death, his wife had his value of π engraved onto his tombstone in St. Pieter's Kerk in Leiden, Holland.

Earlier we mentioned the work of John Wallis (1616–1703). He was a professor of mathematics at Cambridge and Oxford universities, and published a book, *Arithmetica infinitorum* (1655), where he presented the formula for π (actually $\frac{\pi}{2}$, which we then merely double to get π):

$$\frac{\pi}{2} = \frac{2 \times 2}{1 \times 3} \times \frac{4 \times 4}{3 \times 5} \times \frac{6 \times 6}{5 \times 7} \times \frac{8 \times 8}{7 \times 9} \times \cdots \times \frac{2n \times 2n}{(2n-1) \times (2n+1)} \times \cdots$$

This product converges[24] to the value of $\frac{\pi}{2}$. That means its double gets closer and closer to the value of π as the number of terms increases.

---

24. A series converges when it approaches a specific value as a limit. That is, the more terms in the series, the closer it will get to the number to which it converges.

Wallis's results were then transformed into a continued fraction[25] by William Brouncker (ca. 1620–1684)[26] by methods that we are not certain of today. Brouncker obtained the following value of $\frac{4}{\pi}$:

$$\frac{4}{\pi} = 1 + \cfrac{1^2}{2 + \cfrac{3^2}{2 + \cfrac{5^2}{2 + \cfrac{7^2}{2 + \cfrac{9^2}{2 + \cdots}}}}}$$

This procedure to get the value of π is not only tedious but also requires quite a few terms before it gets close to the value of π that we know today.

Still, let's take a look at what this continued fraction can tell us. First, notice that we can maintain the pattern of the above continued fraction by taking further squares of successive odd numbers. To inspect the continued fraction, we look at increasing pieces of the fraction, each time cutting off the rest of the fraction at a plus sign. We call these pieces *convergents*.

The first convergent is 1

The second convergent is

$$1 + \frac{1^2}{2} = \frac{3}{2} = 1.5$$

The third convergent is[27]

$$1 + \cfrac{1^2}{2 + \cfrac{3^2}{2}} = 1 + \cfrac{1}{2 + \cfrac{9}{2}} = 1 + \cfrac{1}{\frac{13}{2}} = 1 + \frac{2}{13} = \frac{15}{13} = 1.15384\overline{615384}6$$

---

25. If you are unfamiliar with continued fractions, then see page 146 for a simple introduction.

26. William Lord Viscount Brouncker (ca. 1620–1684), who found this continued fraction, was cofounder and the first president of the Royal Society (1660).

27. The bar over the digits means that the pattern continues indefinitely.

The fourth convergent is

$$1+\cfrac{1^2}{2+\cfrac{3^2}{2+\cfrac{5^2}{2}}}=1+\cfrac{1}{2+\cfrac{9}{2+\cfrac{25}{2}}}=1+\cfrac{1}{2+\cfrac{9}{2+\cfrac{25}{2}}}=1+\cfrac{1}{2+\cfrac{9}{29}}=1+\cfrac{1}{2+\cfrac{18}{29}}=1+\cfrac{1}{\cfrac{76}{29}}=1+\cfrac{29}{76}=\cfrac{105}{76}=1.38157894736842105263$$

The fifth convergent is

$$\frac{945}{789} \approx 1.19771863117870722433346007604563$$

Since these convergents are approximate values of $\frac{4}{\pi}$, to get these primitive approximations of π, we need to multiply the reciprocal of each convergent by 4. Successively, these values for π are

$$1 \cdot 4 = 4$$

$$\frac{2}{3} \cdot 4 = \frac{8}{3} \approx 2.6667$$

$$\frac{13}{15} \cdot 4 = \frac{52}{15} \approx 3.46667$$

$$\frac{76}{105} \cdot 4 = \frac{304}{105} = 2.8952380$$

$$\frac{789}{945} \cdot 4 = \frac{3,156}{945} = \frac{1,052}{315} = 3.3396825$$

Notice how we are beginning (albeit rather slowly) to sandwich in the true value of π; one value is higher, then one is lower, each time getting closer to the true value:[28] 3.14159265358979.... This, too, was a step closer to the modern methods, even though it didn't achieve the same accuracy as the tedious methods of those who kept constructing regular polygons with an ever-increasing number of sides until they almost "looked" like a circle.

As we mentioned earlier, it took centuries to obtain greater and greater accuracy of the value of π. In 1647 the English mathematician John Wallis designated the ratio of the circumference of a circle to

---

28. Remember, we will never be able to write the true value of π in decimal notation since it is always an approximation. The more decimal places we have, the closer we get to the actual value. Here we give an approximation to fourteen-decimal-place accuracy.

the diameter as $\frac{\pi}{\delta}$, where π probably stood for the periphery (which is *not* what π stands for today!) and δ (delta) stood for the diameter. Later, in 1685, Wallis used π to represent the periphery and a small square, $\square$, to represent his ratio $\frac{4}{3.14149...}$, using 3.14149... this was his approximation of today's π. Gradually, mathematicians approached the more universal use of π for the ratio it represents today.

Our knowledge increased in 1668 when the Scottish mathematician James Gregory (1638–1675) anticipated Germany's greatest mathematician of the seventeenth century, Gottfried Wilhelm Leibniz,[29] by five years when he came up with the following approximation formula for π:

$$\frac{\pi}{4} = 1 - \frac{1}{3} + \frac{1}{5} - \frac{1}{7} + \frac{1}{9} - \frac{1}{11} + \cdots$$

This is a very rough approximation, since the series converges very slowly. It would take one hundred thousand terms to get to a five-place accuracy of π.

The Eighteenth Century—When π Gets Its Name

We now are at about the time when another noteworthy moment in the history of π occurs. In 1706 the English mathematician William Jones (1675–1749), in his book, *Synopsis palmariorum matheseos*,[30] used the symbol π for the first time to actually represent the ratio of the circumference of a circle to its diameter. However, the true popularity of the symbol π to represent this ratio came in 1748, when, as noted earlier, one of mathematics' most prolific contributors, the Swiss mathematician Leonhard Euler (1707–1783), used the symbol π in his book *Introductio in analysin infinitorum* to represent the ratio

---

29. Leibniz was credited as the coinventor of the calculus in modern times.

30. "A New Introduction to Mathematics"

of the circumference of a circle to its diameter. A brilliant mathematician with an uncanny memory and ability to do complex calculations, Euler developed numerous methods for calculating π, some of which approached the true value of π more quickly (that is, in fewer steps) than procedures developed by his predecessors. Here he calculated π to 126-place accuracy. One formula that he used to calculate π was the first in a group of series giving successive powers of π. The series below is particularly interesting, since it is a series created by taking the squares of the terms in a harmonic series.[31]

$$\frac{\pi^2}{6} = \frac{1}{2^2} + \frac{1}{3^2} + \frac{1}{4^2} + \frac{1}{5^2} + \dots \text{ [32]}$$

There are many theorems named after Euler, since he wrote profusely in almost all areas of mathematics, yet the most famous formula (if there actually is one) bearing his name is the relationship that ties together a number of seemingly unrelated concepts. It is $e^{\pi i} + 1 = 0$, where e is the base of the natural logarithms,[33] and i is the imaginary unit of the complex numbers ($i = \sqrt{-1}$). In this formula we have five most important numbers: 0, 1, e, i, and π ![34] This formula prompted the famous German mathematician (Christian) Felix Klein (1849–1925) to proclaim: "All Analysis Lies Here!"

---

31. A harmonic sequence is formed by taking the reciprocals of the terms of an arithmetic sequence (one with a common difference between terms). The simplest arithmetic sequence is 1, 2, 3, 4, 5, 6, ... The related harmonic series is:

$$1 + \frac{1}{2} + \frac{1}{3} + \frac{1}{4} + \frac{1}{5} + \frac{1}{6} \dots$$

The name "harmonic" comes from the fact that a set of strings of the exact same type and with the same torsion, yet of lengths proportional to the terms of a harmonic sequence, when strummed together will produce a harmonic tone.

32. More about this unusual relationship is presented in Herbert Hauptmann's afterword (p. 284).

33. The power to which a base must be raised to equal a given number. For example, given the base 10, the logarithm of 16 is (approximately) 1.2041 because 101.2041 equals (approximately) 16. Both natural logarithms (to the base e, which is approximately 2.71828) and common logarithms (to the base 10) are used in computer programming. The natural logarithm $e = \lim_{n \to \infty} \left(1 + \frac{1}{n}\right)^n = 2.718281828459045....$

34. We discussed this formula in chapter 1 (page 30), although it was given in the form $e^{i\pi} = -1$, and made some mention of its accolades.

Approaching the Nineteenth Century

The question about what kind of number is π began to consume mathematicians. With each attempt to get more place values for π, there was always the hope that a pattern would emerge and that there would be a period of digits repeating. This would have then made π a rational number. This was not to happen. In 1794 the French mathematician Adrien Marie Legendre (1752–1833) wrote a book entitled *Élements de Géométrie* in which he proved that π^2 is irrational. It was the first use of the symbol π in a French book. In 1806 he also proved that π is irrational. We know that Aristotle (384–322 BCE) suspected π was an irrational number. But his speculation lasted more than two millennia before being proved correct.

Although the great German mathematician Carl Friedrich Gauss (1777–1855) also weighed in with calculations of π, he employed Zacharias Dahse (1824–1861), a lightning-fast mental calculator, to assist with his research. Dahse, using the formula

$$\frac{\pi}{4} = \arctan\left(\frac{1}{2}\right) + \arctan\left(\frac{1}{5}\right) + \arctan\left(\frac{1}{8}\right)$$

found π correct to two hundred decimal places.[35] Dahse became a legend with his calculating ability. It is believed that he did these calculations mentally. He was known to be able to multiply in his head two eight-digit numbers in forty-five seconds. Multiplying two forty-digit numbers required forty minutes of mental calculation time, and he was able to mentally multiply two one-hundred-digit numbers in eight hours and forty-five minutes. In fairness to Gauss, it should be said that he, too, was a marvelous calculator. It is believed that Gauss's calculating talent enabled him to see patterns and make many mathematical conjectures that he then later proved, establishing them as theorems.

---

35. This formula was developed by the Viennese mathematician L. K. Schulz von Strassnitzky.

The pursuit of an accurate value for π continued. Some efforts made slight progress by increasing the number of correct decimal places for π, while others claimed to have done so but upon further examination had some errors. In 1847 Thomas Clausen (1801–1855), a German mathematician, calculated π correct to 248 decimal places. Then in 1853 William Rutherford, an Englishman, extended this to 440 decimal places. One of Rutherford's students, William Shanks (1812–1882), extended the value of π to 707 decimal places in 1874. However, there was an error in the 528th place, which was first detected in 1946 with the aid of an electronic computer—using seventy hours of running time! Shanks required fifteen years for his calculation.

Entering the Twentieth Century

As the history of π progresses, we must take note of the work of Carl Louis Ferdinand Lindemann (1852–1939), a German mathematician who proved that π was not only not a rational number, but, in fact, it is a transcendental number.[36] As noted earlier, with the establishment that π was a transcendental number, Lindemann finally put to rest that ancient problem of finding the length of the side of a square whose area is equal to that of a given circle, when he proved that it was impossible to be done.

In chapter 1 we discussed Buffon's needle technique as a method of calculating the value of π. This seemingly unrelated field of probability seemed to relate to π. It is truly astonishing that this geometric ratio, π, would be related to a situation in probability. In the same way, in 1904, R. Chartres showed that the probability that two randomly selected pos-

---

36. A transcendental number is one that cannot be the root of an algebraic equation with rational coefficients. For example, $\sqrt{2}$ is an irrational number but not a transcendental number; it is the root of the equation $x^2 - 2 = 0$. On the other hand, e is a transcendental number (see note 29).

itive integers are relatively prime[37] is $\frac{6}{\pi^2}$. This might be even more amazing, since at least with Buffon's needle there is something physical going on: the placement of a needle and parallel lines. Here there is nothing geometric, just number theory.

In 1914 the Indian mathematical genius Srinivasa Ramanujan (1882–1920),[38] established many formulas for calculating the value of π. Some were very complicated and had to wait for the advent of the computer to be appropriately used. One such is

$$\frac{1}{\pi} = \frac{\sqrt{8}}{9,801} \sum_{n=0}^{\infty} \frac{(4n)!(1,103 + 26,390n)}{(n!)^4 \, 396^{4n}}$$

Yet a much simpler formula that Ramanujan produced to calculate the value of π was

$$\sqrt[4]{9^2 + \frac{19^2}{22}} = \left(81 + \frac{361}{22}\right)^{\frac{1}{4}} = \left(\frac{2,143}{22}\right)^{\frac{1}{4}} = 3.141592652...$$

which is correct to only eight decimal places, but is relatively easy to calculate.[39]

In 1946 D. F. Ferguson (England) discovered an error, as noted earlier, in William Shanks's value of π in the 528th decimal place. In January 1947 he produced a value for π correct to 710 places. Later that month, John W. Wrench Jr., an American, published a value of π to 808 decimal places, but soon thereafter Ferguson found an error in the 723rd decimal place. In January 1948 the two collaborated on a correct value of π to 808 decimal places with the help of a desk calculator. Still using only a desk calculator, the following year John W. Wrench Jr. and Levi B. Smith, American mathematicians, extended this to 1,120 decimal places.

---

37. Two numbers are said to be relatively prime if their only common factor is 1. For example, 15 and 17 are relatively prime, since their only common factor is 1.

38. More about him in chapter 3.

39. All that needs to be done with a simple calculator is to take the square root of the square root of $\frac{2,143}{22}$, that is, $\sqrt{\sqrt{\frac{2,143}{22}}}$.

The Computer Enters the Story of π

In 1949, with the development of the electronic computer, the race for the most decimal place values for π took on a fervor. Now computing time was no longer a factor. We were not limited to human limitations. Using seventy hours of computer time, the brilliant mathematicians John von Neumann, George Reitwiesner, and N. C. Metropolis calculated the value of π to 2,037 decimal places, using an ENIAC computer.

And so the race was on. To inspect each of the methods used is far beyond the scope of this book. Yet we can observe the gradual progress with the help of the following table:[40]

Year	Mathematician	Number of place accuracy of π	Time for calculation
1954	S. C. Nicholson & J. Jeenel	3,092	13 minutes
1954	G. E. Felton	7,480	33 hours
	(Generated 10,021 places but only 7,480 were correct due to machine error.)		
1958	François Genuys	10,000	100 minutes
1959	François Genuys	16,167	4 hours, 20 minutes
1961	Daniel Shanks[41] & John W. Wrench Jr.		
		100,265	8 hours, 43 minutes
1966	M. Jean Guilloud & J. Filliatre	250,000	41 hours, 55 minutes
1967	M. Jean Guilloud & Michele Dichampt		
		500,000	44 hours, 45 minutes
1973	M. Jean Guilloud & Martine Bouyer	1,001,250	23 hours, 18 minutes
1981	Kazunori Miyoshi & Kazuhika Nakayama		
		2,000,036	137 hours, 20 minutes
1982	Yoshiaki Tamura & Yasumasa Kanada		
		8,388,576	6 hours, 48 minutes
1982	Yoshiaki Tamura & Yasumasa Kanada		
		16,777,206	Less than 30 hours
1988	Yoshiaki Tamura & Yasumasa Kanada		
		201,326,551	About 6 hours
1989	Gregory V. & David V. Chudnovsky	1,011,196,691	Not known
1992	Gregory V. & David V. Chudnovsky	2,260,321,336	Not known
1994	Gregory V. & David V. Chudnovsky	4,044,000,000	Not known
1995	Takahashi & Yasumasa Kanada	6,442,450,938	Not known
1997	Takahashi & Yasumasa Kanada	51,539,600,000	About 29 hours
1999	Takahashi & Yasumasa Kanada	206,158,430,000	Not known
2002	Yasumasa Kanada	1,241,100,000,000	About 600 hours

40. For a more complete list of the development of the value of π, see the table on pages 75–77.
41. Daniel Shanks is no relation to William Shanks.

The race for the most number of decimal places for π entered the billions with the Chudnovsky brothers, David and Gregory. Their story is a bit unusual. They emigrated to the United States from the Soviet Union in 1978 after getting doctorates in mathematics from the Ukrainian Academy of Sciences. They took an apartment in Manhattan and rented two supercomputers to do their calculations—bent on getting the most accurate value for π. There were some problems along the way. Gregory, the younger by five years, had myasthenia gravis, an autoimmune disorder of the muscles, and had to stay in bed most of the time. He did most of his work from his bed. Both brothers were married and for a time lived off the earnings of their respective wives, while they pursued their mathematical challenges. The expense of the supercomputers forced them eventually to build their own—taking up much of their apartment. In 1981 things got a bit easier when Gregory won a MacArthur Foundation fellowship in mathematics. This provided much-needed medical insurance and solved their immediate financial problems. Gregory continued to work from his bed, writing mathematical formulas and pursuing the value of π, while also breaking ground in a number of other areas of mathematics. This is just one of many stories to be found in the rich history of π.

There are many unsolved problems in mathematics that beg for solution in addition to the pursuit of π. Perhaps one of the simplest to mention is known as Goldbach's conjecture. It states that any even number greater than 2 can be expressed as the sum of two prime numbers. This conjecture has plagued mathematicians for over 250 years. Despite the fact that using computers we have been able to show that the conjecture holds true for all even numbers so far tested, we have not yet been able to come up with a proof that will show it is true for *all* even numbers greater than 2. In a like way, mathematicians have been driven to try to calculate π to ever-greater accuracy. Of course, from the point of view of usable accuracy, these incredibly long decimal expansions may seem unnecessary. However, as you will later see, there can be a use for these decimal expansions, namely, as a table of random numbers, which can aid in statistical sampling.[42]

As for continuously using computers to establish a greater accuracy for π, it has now gotten to the point where computer scientists are no longer just interested in pushing for greater accuracy for the value of π; rather, they do this to test their computers. How fast, how accurately, and how far can a new computer or computer program calculate the value of π? Mathematicians and π enthusiasts are always looking to extend our knowledge of π. They are interested both in extending the number of known decimal places and in the cleverness of the program or algorithm used to generate these record-breaking attempts. Computer scientists still find the algorithms for the calculation of π ideal tools for testing high-powered supercomputers. So, how far will the next level of accuracy take us in our knowledge of π? And, of course, how much computer time will be required? While these questions plague the computer scientists, π enthusiasts are more interested in the product. Will greater accuracy for the π approximation (now already over 1.24 trillion decimal places) reveal new ideas about π? And will there be more-elegant (and efficient) algorithms discovered for establishing these approximations of π? Both groups of scientists push on, though with different, albeit complementary goals.

---

42. This many not be an ideal table of random numbers since, as we mentioned earlier, the frequency of the digits is not consistent over equal periods.

Here is a summary of the history of the pursuit of the value of π:

Table of computation of Pi from 2000 BCE to the Present

Who calculated π	When	Number of decimal place accuracy	Value found
Babylonians	2000? BCE	1	$3.125 = 3 + 1/8$
Egyptians	2000? BCE	1	$3.16045 \approx 4\left(\dfrac{8}{9}\right)^2$
China	1200? BCE	1	3
Bible (1 Kings 7:23)[43]	550? BCE	1 (4)	3 (3.1416)
Archimedes	250? BCE	3	3.1418
Vitruvius	15 BCE	1	3.125
Hon Han Shu	130 CE	1	$3.1622 \approx \sqrt{10}$
Ptolemy	150	3	3.14166
Wang Fau	250?	1	$3.155555 = \dfrac{142}{45}$
Liu Hui	263	5	3.14159
Siddhanta	380	3	3.1416
Tsu Ch'ung Chi	480?	7	$3.1415926 \approx \dfrac{355}{113}$
Aryabhata	499	4	$3.14156 = \dfrac{62,832}{20,000}$
Brahmagupta	640?	1	$3.162277 \approx \sqrt{10}$
Al-Khowarizmi	800	4	3.1416
Fibonacci	1220	3	3.141818
Al-Kashi	1430	12	3.1415926535898732
Otho	1573	6	3.1415929
Viète	1593	9	3.1415926536
Romanus	1593	15	3.141592653589793
van Ceulen	1596	20	3.14159265358979323846
van Ceulen	1615	35	3.14159265358979323846264338327950029

Newton	1665	16	
			3.1415926535897932
Sharp	1699	71	
Seki Kowa	1700?	10	
Machin	1706	100	
De Lagny	1719	127	(only 112 correct)
Takebe	1723	41	
Kamata	1730?	25	
Matsunaga	1739	50	
Von Vega	1794	140	(only 136 correct)
Rutherford	1824	208	(only 152 correct)
Strassnitzky / Dase	1844	200	
Clausen	1847	248	
Lehmann	1853	261	
Rutherford	1853	440	
William Shanks	1873	707	(only 527 correct)
Ferguson	1946	620	
Ferguson	Jan. 1947	710	
Ferguson and Wrench	Sep. 1947	808	
Smith and Wrench	1949	1,120	
Reitwiesner et al. (ENIAC)	1949	2,037	
Nicholson and Jeenel	1954	3,092	
Felton	1957	7,480	
Genuys	Jan. 1958	10,000	
Felton	May 1958	10,021	
Genuys	1959	16,167	
Daniel Shanks and Wrench	1961	100,265	
Guilloud and Filliatre	1966	250,000	
Guilloud and Dichampt	1967	500,000	
Guilloud and Bouyer	1973	1,001,250	
Miyoshi and Kanada	1981	2,000,036	
Guilloud	1982	2,000,050	

Tamura	1982	2,097,144
Tamura and Kanada	1982	4,194,288
Tamura and Kanada	1982	8,388,576
Kanada, Yoshino, and Tamura	1982	16,777,206
Ushiro and Kanada	Oct. 1983	10,013,395
Gosper	Oct. 1985	17,526,200
Bailey	Jan. 1986	29,360,111
Kanada and Tamura	Sep. 1986	33,554,414
Kanada and Tamura	Oct. 1986	67,108,839
Kanada, Tamura, Kubo et al.	Jan. 1987	134,217,700
Kanada and Tamura	Jan. 1988	201,326,551
Chudnovskys	May 1989	480,000,000
Chudnovskys	Jun. 1989	525,229,270
Kanada and Tamura	Jul. 1989	536,870,898
Chudnovskys	Aug. 1989	1,011,196,691
Kanada and Tamura	Nov. 1989	1,073,740,799
Chudnovskys	Aug. 1991	2,260,000,000
Chudnovskys	May 1994	4,044,000,000
Takahashi and Kanada	Jun. 1995	3,221,225,466
Takahashi and Kanada	Aug. 1995	4,294,967,286
Takahashi and Kanada	Sep. 1995	6,442,450,938
Takahashi and Kanada	Jun. 1997	51,539,600,000
Takahashi and Kanada	Apr. 1999	68,719,470,000
Takahashi and Kanada	Sep. 1999	206,158,430,000
Kanada and nine-person team at University of Tokyo		
	Sep. 2002	1,241,100,000,000

---

43. Using gematria—see chapter 1.

Chapter 3

Calculating
the Value of π?

Up to now we have described π and mentioned ways in which attempts have been made to calculate its value. They varied from highly intelligent (say, ingenious) guesses by mathematicians, to attempts at performing calculations that were later proved impossible (i.e., squaring the circle), to carefully planned constructions that would yield the value of π if carried out far enough and carefully enough. Some methods of calculating the value of π, strangely enough, relied on probability, or in one case on mysterious insights. Here we will provide you with a variety of methods for calculating the value of π. We chose those that should be easily understood by the general reader. Where a concept is used that may be a bit off the beaten path, or simply unfamiliar to some, we provide some background information. We will be presenting the classical attempts, rather than those used in the more recent computer-driven methods. We begin with one of the most famous classical methods, by one of the most gifted mathematicians in the history of mathematics, Archimedes.

Archimedes' Method for Finding the Value of π

Perhaps the easiest way to begin to calculate the value of π was developed by Archimedes. It is a method that can appeal to one's intuition. He noticed that as the number of sides of a regular polygon increases, while keeping the radius or the apothem[1] constant, the limiting value of the perimeter is the circumference of a circle. That is, suppose we take the first few regular polygons (an equilateral triangle, a square, a regular pentagon, and a regular hexagon) and inscribe them in the same-size circle. As the number of sides of the regular polygon increases, the perimeter of the polygon gets closer and closer to the circumference (i.e., perimeter) of the circle. Remember that the circumscribed circle must contain each of the vertices of the polygon. Here is what it can look like.

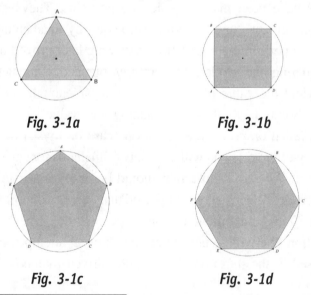

Fig. 3-1a   Fig. 3-1b

Fig. 3-1c   Fig. 3-1d

---

1. The apothem is the segment from the center of a regular polygon to the midpoint of one of its sides. It is perpendicular to the side.

This may be easier to see when the regular polygon's sides increase further so that it becomes a dodecagon (which has twelve sides). We can actually calculate the increasing perimeters and see them gradually approach the circumference of the circle.

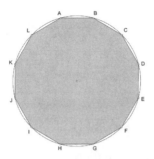

Fig. 3-2

Let's take the hexagon as our example of a "general polygon." From this we will then generalize to polygons of many more (or fewer) sides. We begin with a regular hexagon inscribed in a circle of radius $\frac{1}{2}$. The measure of $\angle AOB$ is one-sixth of a complete revolution of 360°, or 60°. Since $\overline{OK} \perp \overline{AB}$ at K, $BK = AK = a$.

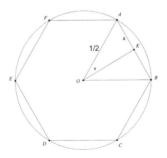

Fig. 3-3

We seek to find the perimeter of the hexagon, when we know the length of the radius $\left(\frac{1}{2}\right)$ and the measure of

$$\angle AOK = \frac{1}{2}(60°) = 30°$$

Using the trigonometric function sine,[2] we get

$$\angle AOK = \sin 30° = \frac{a}{\frac{1}{2}} = 2a$$

Since $\sin 30° = \frac{1}{2}$, then $2a = \frac{1}{2}$, and $a = \frac{1}{4}$. The perimeter of the hexagon is then 12 times a, which equals 3.

Let's generalize this for any regular polygon of n sides.

$$\angle x = \frac{1}{2} \cdot \frac{360°}{n} = \frac{180°}{n}$$

Therefore, for the general regular polygon of n sides

$$\sin \frac{180°}{n} = 2a$$

The perimeter of the n-sided regular polygon is then n times $2a$, which makes this perimeter equal to

$$n \sin \frac{180°}{n}$$

We can then take various values of n and compute the perimeter of the regular polygon whose circumscribed circle has a radius of $\frac{1}{2}$.

---

2. The sine function is defined for a right triangle as the ratio of the side opposite the angle in question and the hypotenuse (the side opposite the right angle).

We shall work out the first few examples here and then provide the results of others in a table.

When $n = 3$:

$$3\sin\frac{180°}{3} = 3\sin 60° \approx 3(0.8660254037844386467637231 7075294)$$
$$= 2.5980762113533159402911695122588$$

When $n = 4$:

$$4\sin\frac{180°}{4} = 4\sin 45° \approx 4\left(0.707106781186547524400844362 10485\right)$$
$$= 2.8284271247461900976033774484194$$

When $n = 5$:

$$5\sin\frac{180°}{5} = 5\sin 36° \approx 5\left(0.58778525229247312916870595463907\right)$$
$$= 2.9389262614623656458435297731954$$

When $n = 6$:

$$6\sin\frac{180°}{6} = 6\sin 30° = 6\left(0.50000000000000000000000000000000\right)$$
$$= 3.00000000000000000000000000000000$$

We just calculated the first four entries in the table below and here provide you with the remaining ones.

n	Perimeter of inscribed polygon of n sides
3	2.598076211353315940291169512258...
4	2.828427124746190097603377448419...
5	2.938926261462365645843529773195...
6	3.000000000000000000000000000000...
7	3.037186173822906843330378329938...
8	3.061467458920718173827679872243...
9	3.078181289931018597396896532140...
10	3.090169943749474241022934171828...
11	3.099058125255726674825597068812...
12	3.105828541230249148186786051488...
13	3.111103635738250972933798441382...
14	3.115293075388401660044635902955...
15	3.118675362266390056526134266076...
24	3.132628613281238197161749469491...
36	3.137606738915694248090313750149...
54	3.139820761165694741092392909741...
72	3.140595890304191984286221559116...
90	3.140954703225087448139566346280...
120	3.141233796944778313273402266493...
180	3.141433158711032307495416132936...
250	3.141509970838151978568647287198...
500	3.141571982779475624867655078979...
1,000	3.141587485879563351933227035495...
10,000	3.141592601912665692979346479289...

Now compare this 10,000-sided regular polygon (our last entry) to the value of π that we already know. Remember it is inscribed in the circle with radius $\frac{1}{2}$. This 10,000-sided polygon is optically quite indistinguishable from the circle (obviously, without magnification enhancements). The circumference of the circumscribed circle of radius $\frac{1}{2}$ is $2\pi r = 2\pi\left(\frac{1}{2}\right) = \pi$.

Look at the known value of π for comparison.

π ≈ 3.1415926535 8979323846 2643383279 5028841971 6939937510 5820974944…

Up to the seventh decimal place, the approximation with a 10,000-sided regular polygon perimeter is correct. If we were to calculate the perimeter of a regular polygon of 100,000 sides, we would get an even closer approximation. The perimeter of a regular polygon of 100,000 sides is 3.14159265307302196048314802075331…, which approximates π correct to nine decimal places.

Archimedes (obviously) did not have the luxury of using electronic (or even mechanical) calculating devices to assist him in his calculations.[3] He also did not have the facility brought about by the place value system (such as our decimal system), nor did he have the use of trigonometry available to him. Yet he still used a 96-sided regular polygon. He saw the circle as the limiting figure of an inscribed circle as well as the circumcircle we just used above. By taking the average of the perimeters of each pair of circles of n-sided regular polygons, he would "sandwich in" the perimeter of the circle, which in the case of a circle with radius of $\frac{1}{2}$ is π.

---

3. The mechanical calculator was invented by four mathematicians over a rather wide stretch of time. Wilhelm Schickardt (1592–1635), a German mathematician, built the first digital calculator in 1623. Blaise Pascaal (1623–1662) built the first mechanical calculating machine in 1642 for his father, who was a tax collector. The machine, called Pascaline, was commercially sold after 1645. Gottfried Wilhelm Leibniz (1646–1716) developed a mechanical calculator in 1673 that failed during a demonstration in London but nonetheless, because of the spectacular concept involved, was accepted in the Royal Society. The English mathematician Charles Babbage (1792–1871), despite devoting a greater part of his professional life to its development, never reached a completed product.

Let us now repeat the above exercise with the polygon circumscribed about the circle, or, put another way, where the circle of radius $\frac{1}{2}$ is inscribed in the polygon (i.e., the circle must be tangent to each side of the polygon). As before, we will consider regular polygons with successively greater numbers of sides, each with our given circle inscribed.

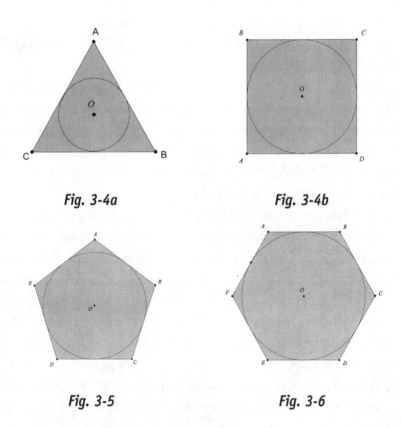

Fig. 3-4a   Fig. 3-4b

Fig. 3-5   Fig. 3-6

Notice how gradually the perimeter of the polygon appears to get closer and closer to the circumference of the circle.

He began his work in 1812 and worked for decades on the project. In the end, the lack of precision tools prevented him from achieving his "analytical engine" in 1833. His work was first realized in the form of a working machine in 1944 when the IBM Corporation and Harvard University collaborated to produce the Automatic Sequence Controlled Calculator.

This time we will consider a regular pentagon circumscribed about our circle of radius $\frac{1}{2}$ as our first polygon to study. Then we will generalize our procedure and extend it to many others.

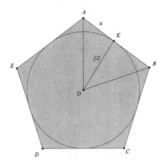

Fig. 3-7

Our objective is to find the perimeter of the pentagon with a side of $2a$. We know that

$$\tan \angle AOK = \frac{a}{OK} \text{ }^{4} \text{ and } m \angle AOB = 72°$$

so that

$$m\angle AOK = 36°, \text{ while } OK = \frac{1}{2}$$

Therefore,

$$a = \frac{1}{2}\tan 36° \approx \frac{1}{2}(0.726542528005360885895466675748062)$$
$$= 0.363271264002680442947733337874031$$

Thus the perimeter of the pentagon is 10 times a, or about 3.632712640026804429477333787403I (this is found by taking 5 times $2a$, or about 3.6327), not yet a very close approximation of π. The circumference of the circle is $2\pi r = 2\pi\left(\frac{1}{2}\right) = \pi$.

In the general case of a regular polygon of n sides

$$m\angle AOK = \frac{1}{2} \cdot \frac{360°}{n} = \frac{180°}{n}$$

From the example of the pentagon, $\tan \angle AOK = \frac{a}{OK}$. It follows that

$$a = OK \tan \angle AOK = \frac{1}{2} \cdot \tan \frac{180°}{n}$$

The perimeter of the polygon is then

$$n \cdot 2a = n \cdot 2 \cdot \frac{1}{2} \cdot \tan \frac{180°}{n} = n \tan \frac{180°}{n}$$

As before, we will calculate the perimeters of the various polygons, this time, though, circumscribed about our circle with radius $\frac{1}{2}$.

We already have the calculated perimeter for the pentagon, so we will do the calculation for the hexagon now.

When $n = 6$:

$$n \cdot \tan \frac{180°}{n} = 6 \cdot \tan \frac{180°}{6} \cdot \tan 30° = 6 \cdot \frac{\sqrt{3}}{3} \approx 3.4641$$

For more than four decimal places we get the following:

n	Perimeter of circumscribed polygon of n sides
3	5.19615242270663188058233902451 76...
4	4.00000000000000000000000000000000...
5	3.63271264002680442947733378740 31...
6	3.46410161513775458705489268301 17...
7	3.37102233165270051032513647139 88...
8	3.31370849898476039041350979367 76...
9	3.27573210839582125215943094499 15...
10	3.24919696232906326155871412215 13...
11	3.22989142232203385420668296859 44...
12	3.21539030917347247767064390192 95...
13	3.20421221941570764730031492162 91...
14	3.19540864146209913308655906885 42...
15	3.18834842505033187889387490855 12...
24	3.15965994209750048331663497783 32...
36	3.14959188693326418799267209965 86...
54	3.14514184337910393914934210860 04...
72	3.14358788941286845956260303991 74...
90	3.14286925425729574503623631963 53...
96	3.14271459964536829816885909377 21...
120	3.14231058830243146672365927534 28...
180	3.14191168707916543772320113955 1...
250	3.14175803084489443537076906133 84...
500	3.14163399594488606459529576947 32...
1,000	3.14160298905615612604134329010 54...
10,000	3.14159275694405291972467077191 18...

Again you will notice how the more sides the polygon has, the closer its perimeter gets to the circumference of the circle—which we now know is π.

Archimedes, as we said before, saw the inscribed and the circumscribed polygons "sandwiching in" the circle, as seen below by the inscribed and circumscribed dodecagons (n = 12).

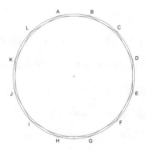

Fig. 3-8

He essentially suggested taking the average of the two perimeters for each type of polygon to get a better approximation of π.

n	Perimeter of inscribed polygon of n sides	Perimeter of circumscribed polygon of n sides	Average of the perimeters of the inscribed and circumscribed polygons of n sides
3	2.5980762113533159402911695122588...	5.1961524227066318805823390245176...	3.8971143170299739104367542683875...
4	2.8284271247461900976033774484194...	4.0000000000000000000000000000000...	3.4142135623730950488016887242095...
5	2.9389262614623656458435297731954...	3.6327126400268044294773337874031...	3.2858194507445850376604317800299...
6	3.0000000000000000000000000000000...	3.4641016151377545870548926830117...	3.2320508075688772935274463415055...
7	3.0371861738229068433303783299385...	3.3710223316527005103251364713988...	3.2041042527378036768277574006668...
8	3.0614674589207181738276798722432...	3.3137084989847603904135097936776...	3.1875879789527392821205948329 6...
9	3.0781812899310185973968965321403...	3.2757321083958212521594309449915...	3.1769566991634199247781637385655...
10	3.0901699437494742410229341718282...	3.2491969623290632615587141221513...	3.1696834530392687512908241469895...
11	3.0990581252557266748255970688128...	3.2298914223220338542066829685944...	3.1644747737388880264516140018703...
12	3.1058285412302491481867860514886...	3.2153903091734724776706439019295...	3.1606094252018608192871497670 85...
13	3.1111036357382509729337984413828...	3.2042122194157076473003149216291...	3.1576579275769793101170566815055...
14	3.1152930753884016600446359029551...	3.1954086414620991330865590688542...	3.1553508584252503965655974859045...
15	3.1186753622663900565261342660769...	3.1883484250503318788938749085512...	3.1535118936583609677100045873135...
24	3.1326286132812381971617494694917...	3.1596599420975004833166349778332...	3.1461442776893693402391922 3662...
36	3.1376067389156942480903137 50149...	3.1495918869332641879926720996586...	3.1435993129244792180414929249035...
54	3.1398207611656947410923929097419...	3.1451418433791039391493421086004...	3.1424813022723993401208675091705...
72	3.1405958903041919842862215591 16...	3.1435878894128684595626030399174...	3.1420918898585302219244122995165...
90	3.1409547032250874481395663 4628...	3.1428692542572957450362363196353...	3.1419119787411 91596587901332 9575...
96	3.1410319508905096381113529264597...	3.1427145996453682981688590937721...	3.1418732752679389681401060101155...
120	3.1412337969447783132734022664935...	3.1423105883024314667236592753428...	3.1417721926236048899985307709175...
180	3.1414331587110323074954161329369...	3.1419116870791654377232011395 51...	3.1416724228950988726093086362435...
250	3.1415099708381519785686472871987...	3.1417580308448944353707690613384...	3.1416340008415232069697081742 68...
500	3.1415719827794756248676550789799...	3.1416339959448860645952957694732...	3.1416029893621808447314754242 26...
1,000	3.1415874858795633519332270354959...	3.1416029890561561260413432901054...	3.1415952374678597389872851628...
10,000	3.1415926019126656929793466 79289...	3.1415927569440529197246707719118...	3.1415926794283593063520086256...

The average of the two perimeters (the right-hand column) is closest to the value of π for each type of polygon. When Archimedes did these calculations, he didn't take as many examples as we did here. He began with two regular hexagons, then doubled the number of sides, using two dodecagons (12-sided polygons), then used two 24-gons,[5] then two 48-gons, and then two 96-gons.

Although his calculations were probably not as accurate as ours are, and we do not have a record of how he did his calculations, he did conclude from the 96-gon that the ratio of the circumference of a circle to its diameter—which is π—is greater than $3\frac{10}{71}$ and less than $3\frac{1}{7}$. We can write this symbolically as

$$3\frac{10}{71} < \pi < 3\frac{1}{7}$$

For comparison to the above, this is

$$3.140845070422535211267605633802 8\ldots < \pi <$$
$$3.142857142857142857142857142857 1\ldots$$

We have come a long way since Archimedes' ingenious methods. As we noted earlier, we can now calculate π to many more places than ever thought possible; however, this "primitive" method gives much intuitive insight into what this ratio that π represents really is.

A Reverse Method to Archimedes by Cusanus

Archimedes had used inscribed and circumscribed regular polygons within and about a given circle, each time increasing the number of sides. The argument was that as the number of sides of the polygons

---

5. 24-gon is a short way of referring to a 24-sided polygon.

increased, the circumference of the circle, "sandwiched" between the two polygons, was the limiting value of the polygon.

An analogous method developed by Nicholas of Cusa (1401–1464) has us "sandwiching" in regular polygons with increasing numbers of sides by inscribed and circumscribed circles. Nicholas of Cusa[6] took his name from his hometown of Cues (today Kues) on the Mosel River in Germany. By today's assessments, he is considered one of the pioneering German thinkers in the transition from the Middle Ages to modern times, yet he was not too well known as a mathematician. He was better known for his substantial career in the church. He became a cardinal in 1488 and was bishop of Brixen (northern Italy) and a governor (or vicar general) of Rome. As a mathematician, he made ill-fated attempts to square the circle[7] and to trisect the general angle,[8] both of which we now know are impossible. As with many mathematicians fascinated with one of the three "famous problems of antiquity," namely, squaring the circle, Cusanus's attempts led him to a fine approximation of π. Let's take a look at what Cusanus achieved in these attempts. We will demonstrate this here, but in more modern terms.

In 1450 Cusanus nested a given regular polygon with the fixed perimeter 2 with inscribed and circumscribed circles. He used a sequence of the regular n-gons ($n = 4, 8, 16, 32, ...$).

---

6. Sometimes referred to by his Latin name, Cusanus.

7. One of the three famous problems of antiquity—no longer a problem today—is how to "square a circle." That means how to construct (with only an unmarked straightedge and a pair of compasses) the side length of a square equal in area to a given circle. Today, we know this to be impossible.

8. Another of the three famous problems of antiquity—no longer a problem today—is how to "trisect an angle." That means how to construct (with only an unmarked straightedge and a pair of compasses) angle trisectors of a general angle—not any specific number of degrees, for it would be possible for some special angles, such as a right angle. Today, we know this general angle trisection to be impossible.

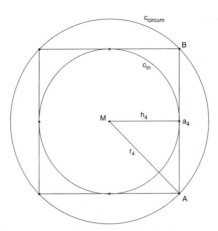

Fig. 3-9

Let's begin as Cusanus did by starting with a square (which may, of course, be referred to as a regular 4-gon). We will call the perimeter of the square p_4. Since each side of the square is $a_4 = \frac{1}{2}$, then $p_4 = 4 \cdot a_4 = 2$.

Consider the inscribed circle with circumference C_{in} and the circumscribed circle with circumference C_{circum} of the square pictured above. The radius of the inscribed circle is

$$h_4 = \frac{a_4}{2} = \frac{1}{4} = 0.25$$

The radius of the circumscribed circle is

$$r_4 = \sqrt{h_4^2 + \left(\frac{a_4}{2}\right)^2} = \frac{\sqrt{2}}{4} \approx 0.3535533905$$

We can plainly see that the perimeter of the square is somewhere between the circumferences of the two circles, so we can compare the perimeter, p_4, and the two circumferences, C_{in} and C_{circum}, to get

$$C_{in} < p_4 < C_{circum} \quad \text{or} \quad 2\pi h_4 < 2 < 2\pi r_4$$

Dividing all terms by 2 gives us

$$\pi h_4 < 1 < \pi r_4$$

Then dividing all terms by π we get

$$h_4 < \frac{1}{\pi} < r_4$$

Taking the reciprocal values of each of the terms reverses the inequality, so we get

$$\frac{1}{r_4} < \pi < \frac{1}{h_4}$$

Since $r_4 = \frac{\sqrt{2}}{4}$, it follows that $\frac{1}{r_4} = \frac{4}{\sqrt{2}} \approx 2.82842713$ and $\frac{1}{h_4} = 4$. Therefore $2.82842713 < \pi < 4$, a rather rough approximation for the value of π. But wait, as we increase the number of sides of the regular polygon, the estimates should get better.

The next approximation was done by doubling the number of sides of the regular polygon we just used to get a regular octagon. Cusanus considered a regular 8-gon with h_8 as the radius of the inscribed circle with circumference C_{in} and r_8 as the radius of the circumscribed circle with circumference C_{circum}.

Since each side is $a_8 = \frac{1}{4}$, the perimeter, p_8, of the regular 8-gon is $p_8 = 8 \cdot a_8 = 2$.

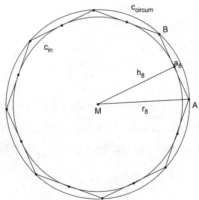

Fig. 3-10

$m\angle AMB = 45°$, therefore, $m\angle AMC = 22.5°$

$$\tan \angle AMC = \tan 22.5° = \sqrt{2} - 1 = \dfrac{\dfrac{a_8}{2}}{h_8} = \dfrac{a_8}{2h_8} \quad 9$$

This can then be transformed into the following equation:

$$h_8 = \dfrac{a_8}{2\tan\angle AMC} = \dfrac{\dfrac{1}{4}}{2 \cdot \left(\sqrt{2}-1\right)} = \dfrac{1}{8\left(\sqrt{2}-1\right)} = \dfrac{\sqrt{2}+1}{8} \approx 0.3017766952 \quad 10$$

With the Pythagorean theorem, $r_8^2 = h_8^2 + \left(\dfrac{a_8}{2}\right)^2$, we get

$$r_8 = \sqrt{\dfrac{\sqrt{2}}{32} + \dfrac{1}{16}} \approx 0.3266407412$$

For the perimeter, p_8, and circumferences, C_{in} and C_{circum}, we get

$$C_{in} < p_8 < C_{circum} \quad \text{or} \quad 2\pi \, h_8 < 2 < 2\pi \, r_8$$

Dividing both sides by 2, it follows that $\pi h_8 < 1 < \pi r_8$, which, when dividing each term by π, yields $h_8 < \dfrac{1}{\pi} < r_8$.

---

9. Tan 22.5° = $\sqrt{2}$ – 1 can be obtained by applying a theorem from the high school course to an isosceles right triangle with one of its base angles bisected to get 22.5°.

The theorem states that the angle bisector of a triangle divides the side to which it is drawn proportionally to the two adjacent sides. Hence,

$\dfrac{\sqrt{2}}{1} = \dfrac{1-x}{x}$, then $x\sqrt{2} = 1 - x$, and $x = \dfrac{1}{\sqrt{2}+1}$

Rationalizing the denominator gives us

$x = \dfrac{1}{\sqrt{2}+1} \bullet \dfrac{\sqrt{2}-1}{\sqrt{2}-1} = \sqrt{2} - 1$

10. This was obtained by rationalizing the denominator, that is, by multiplying by 1 in the form of $\dfrac{\sqrt{2}+1}{\sqrt{2}+1}$.

Again, taking the reciprocal of each term reverses the inequality to give us

$$\frac{1}{r_8} < \pi < \frac{1}{h_8}$$

That means for the reciprocal values

$$\frac{1}{r_8} = \sqrt{32 - 16\sqrt{2}}$$

$$= 4\sqrt{2 - \sqrt{2}} \approx 3.061467458$$

and $\frac{1}{h_8} = 8\,(\sqrt{2} - 1) \approx 3.313708498$

We finally have a more accurate value range for π:

3.061467458 < π < 3.313708498

We will now take a giant leap to the general case, where we will try to sandwich in the value of π. This may be a bit complicated for the reader no longer familiar with some of the intricacies of high school mathematics, yet it is the conclusion of this generalization that is of greater importance than the process.

For the general case, h_n is the radius of the inscribed circle with circumference C_{in}, and r_n is the radius of the circumscribed circle with circumference C_{circum} of a regular n-gon ($n = 4, 8, 16, 32, ...$). We have established above that

$$\frac{1}{r_n} < \pi < \frac{1}{h_n}$$

In this way one can generalize nested intervals for π, if one iteratively[11] determines the radii of the inscribed and circumscribed circles with increasing numbers of sides of the regular n-

---

11. Iteration is a computational procedure in which the desired result is approached through a repeated cycle of operations, each of which more closely approximates the desired result.

gon (with the perimeter 2).

How did Cusanus get his iteration method? To explain this, we look again at a regular n-gon (n = 4, 8, 16, 32, ...):

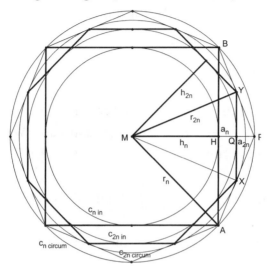

Fig. 3-11

We assume $AB = a_n$, $MA = MB = r_n$, and $MH = h_n$. After doubling the number of sides of the polygon, we get the regular $2n$-gon. Here P is the midpoint of the arc $\overset{\frown}{AB}$, and X and Y are the midpoints of the sides $\overline{AP}$ and $\overline{BP}$ in the triangle $\triangle ABP$. Therefore, $XY = \frac{AB}{2}$. $\overline{XY}$ is the side of the regular $2n$-gon with the perimeter 2 and the center M. It follows that $MP = MA = r_n$, $MX = MY = r_{2n}$, and $MQ = h_{2n}$ (compare the cases where n = 4 and $2n$ = 8 in the above figure).

Because Q is the midpoint of the segment $\overline{PH}$, we have $h_{2n} = \frac{r_n + h_n}{2}$.

In the right triangle ΔMPX it follows that $MX^2 = MQ \cdot MP$.[12] This may be written as $r_{2n}^2 = r_n \cdot h_{2n}$, which then leads to

$$r_{2n} = \sqrt{r_n \cdot h_{2n}}$$

To generate the values for the rest of the n-gons (where $n = 16, 32, 64$, etc.), we can use the general case. We use the following general terms:

$$h_4 = \frac{1}{4}; r_4 = \frac{\sqrt{2}}{4} \qquad \text{(start values)}$$

$$h_{2n} = \frac{r_n + h_n}{2}; r_{2n} = \sqrt{r_n \cdot h_{2n}} \quad \text{(iteration terms)}$$

These yield the following table of values:

n	h_n	r_n	$\frac{1}{r_n}$	$\frac{1}{h_n}$
4	0.25	0.3535533905	2.828427124	4
8	0.3017766952	0.3266407412	3.061467458	3.313708498
16	0.3142087182	0.3203644309	3.121445152	3.182597878
32	0.3172865746	0.3188217886	3.136548490	3.151724907
64	0.3180541816	0.3184377538	3.140331156	3.144118385
128	0.3182459677	0.3183418463	3.141277250	3.142223629
256	0.3182939070	0.3183178758	3.141513801	3.141750369
512	0.3183058914	0.3183118835	3.141572940	3.141632080
1024	0.3183088874	0.3183103855	3.141587725	3.141602510
2048	0.3183096365	0.3183100110	3.141591421	3.141595117
4096	0.3183098237	0.3183099173	3.141592345	3.141593269
8192	0.3183098705	0.3183098939	3.141592576	3.141592807
16384	0.3183098822	0.3183098881	3.141592634	3.141592692
32768	0.3183098852	0.3183098866	3.141592648	3.141592663

We have achieved seven decimal places accuracy for the value of π: 3.1415926. Here for puposes of comparison is the value of π correct to thirty-one decimal places:

π = 3.1415926535897932384626433832795...

---

12. This comes from similar triangles MXP and MQX, or by applying the familiar mean proportional theorems.

While the method of Archimedes relies on the trigonometric functions sine and tangent, the method of Cusanus depends only on elementary theorems like the Pythagorean theorem, similarity, and the basic definition of the trigonometric functions. Furthermore, the arithmetic and geometric means are used for the iteration:

$$A(x,y) = \frac{x+y}{2} = \frac{r_n + h_n}{2} = h_{2n}$$

$$G(x,y) = \sqrt{x \cdot y} = \sqrt{r_n \cdot h_{2n}} = r_{2n}$$

Calculation of the Value of π by Counting Squares

It is always challenging to determine the actual value of π. There isn't any arithmetically comfortable method for calculating the value of π. On the one hand, we only need an elementary knowledge of mathematics for the following methods, but, on the other hand, these approximations of the value of π are not as exact as the calculations done by Archimedes or Cusanus. We offer here a few relatively simple methods for calculating the value of π.

To determine the area of a circle, we can cover it with a lattice of squares (each with a side of length 1), and we will count the number of squares (*a*) in the interior of the circle. We then will count the number of squares (*b*) that are intersected by the circle's circumference.

We will *assume*[13] that one-half of the area of these intersected squares lies in the interior of the circle, and the other half of the area of these intersected squares lies outside the circle.

---

13. It is this assumption that will limit our accuracy of the approximation of π.

So we have $a + \frac{b}{2}$ as an approximate value for the area of the circle. Let's consider this with the following example.

Example 1:
Circle with the radius $r = 8$

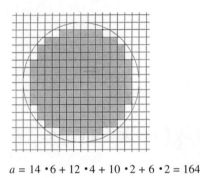

$a = 14 \cdot 6 + 12 \cdot 4 + 10 \cdot 2 + 6 \cdot 2 = 164$ $b = 60$

Fig. 3-12

By the known method (using the formula for the area of a circle) we get

$Area_{circle} = \pi \cdot r^2 \approx 3.14 \cdot 8^2 = 200.96$

Now using the counting method, we get the following result, which compares favorably with the traditional method above.

Approximate value: $Area_{circle} \approx a + \frac{b}{2} = 164 + 30 = 194$, which is close to the "actual" 201 found by the formula.

This approximation leads to π as follows: The approximate area of the circle is 194. This should then be equal to $8^2 \pi = 64\pi$, so that $64\pi \approx 194$. This gives a value for $\pi = \frac{194}{64} \approx 3.03125$.

The approximation of π becomes better when a larger number of squares is used, say, when $r = 10$:

Example 2:

Circle with the radius $r = 10$

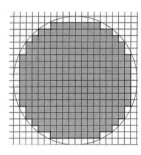 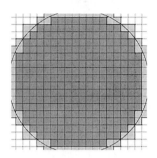

$a = 18 \cdot 8 + 16 \cdot 4 + 14 \cdot 2 + 12 \cdot 2 + 8 \cdot 2 = 276$ $b = 68$

Fig. 3-13

$Area_{circle} = \pi \cdot r^2 \approx 3.14 \cdot 10^2 = 314$

Approximate value: $Area_{circle} \approx a + \frac{b}{2} = 276 + 34 = 310$,

which is now closer to the "actual" value, 314, as determined by the formula above. Again, $Area_{circle} = 310 = 10^2\pi = 100\pi$. Thus, $\pi \approx \frac{310}{100}$ = 3.1, which compares favorably to the previous approximation.

Instead of the whole circle, it suffices to look at a quadrant, count the respective squares, and then quadruple it.

Calculating the Value of π by Counting Lattice Points

The method used by the German mathematician Carl Friedrich Gauss (1777–1855) is relatively simpler. Instead of counting the number of squares, he determined the area of a circle by counting the number of lattice points of the square lattice in the interior of the circle. Lattice points are points with integer coordinates. We can locate all the lattice points of the circle with radius r with $x^2 + y^2 \le r^2$ (the Pythagorean theorem).

If $f(r)$ represents the number of lattice points that lie in the circular area with the radius r, then we get (with the help of Gauss's idea) an approximation for π:

$$\pi \approx \frac{f(r)}{r^2}$$

This should bring back thoughts (especially if you were to multiply both side of this "almost-equation" by r^2) of the now-famous formula for the area of a circle, $Area = \pi r^2$. Here we have $\pi r^2 \approx f(r)$.

There is a formula for finding $f(r)$, but this formula is complicated. Instead, we will give some examples. Consider the following.

Example:

$$f(r) = 317, \, r^2 = 100, \text{ therefore, } \frac{f(r)}{r^2} = 3.17 \approx \pi$$

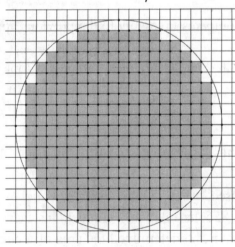

Fig. 3-14

Instead of the whole circle, it suffices to look at a quadrant and to count the respective lattice points again. (Be careful to count the origin only once.)

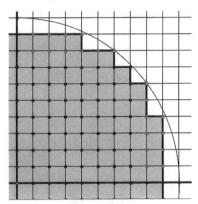

Fig. 3-15

Further values:

r	5	10	20	30	100	200	300	...	100,000
f(r)	81	317	1,257	2,821	31,417	125,629	282,697		31,415,925,457
$\frac{f(r)}{r^2}$	3.24	3.17	3.1425	3.134	3.1417	3.140725	3.14107		3.1415925457

It appears that this sequence heads for the actual value of π, 3.1415926....

For $r = 20$, we already get the correct second place after the decimal point. Strangely enough, for $r = 30$ the value of π gets less accurate, but then eventually gets closer to the true value of π.

Using Physical Properties to Calculate the Value of π

A physicist might determine the value of π by using what may be considered a simpler method than was involved with the tedious task of counting squares or lattice points.

He would weigh (as exactly as possible) one circle with a radius of 10 cm, which he would cut out of an evenly thick piece of cardboard. Then he would compare its weight (or its mass) with that of a square (of 10 cm length side) cut out of the same material.

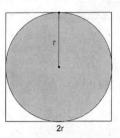

2r

Fig. 3-16

We now compare the area of the circle, radius r, to the area of the square, side $2r$, by considering their weights.[14]

$$\frac{m_{circle}}{m_{square}} = \frac{(2r)^2}{\pi r^2} = \frac{4r^2}{\pi r^2} = \frac{4}{\pi} \approx 1.273239.$$ [15] Therefore, $\pi \approx 3.141594$.

In the eighteenth century, the French agronomist Franzose Olivier de Serres "proved," by using a scale, that a circle weighs as much as a square whose side has the length of that of an equilateral triangle inscribed in the circle—this assumes that both figures are cut out of the same material. When you follow the discussion below, you will see that this implies that $\pi = 3$!

---

14. Technically speaking, we actually have a cylinder and a rectangular solid, if one were to consider the width of the cardboard as the height of these objects.

15. Actually a constant value.

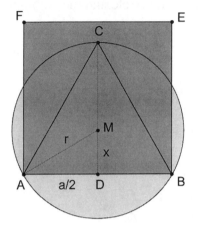

in $\triangle ADM$: $m\angle DAM = \frac{1}{2}(60°) = 30°$

$\sin\angle DAM = \frac{x}{r} = \frac{1}{2}$; therefore, $x = \frac{r}{2}$

$\left(\frac{a}{2}\right)^2 = r^2 - x^2 = \frac{3}{4}r^2$; therefore, $\frac{a}{2} = \frac{\sqrt{3}}{2}$

Area of the square $= a^2 = \left(r\sqrt{3}\right)^2 = \mathbf{3}\ r^2$

Area of the circle $= \pi\ r^2$

We can then conclude that $\pi = 3$.

Fig. 3-17

The Monte-Carlo Method to Determine the Value of π

The Monte-Carlo method[16] is a procedure that makes use of probability, calculus, and statistics to form a summary to establish facts with a large number of tests for a random experiment. The Buffon's needle problem (see chapter 1, pages 38–39) is considered one of the Monte-Carlo methods.

Another such procedure can be simulated by means of raindrops that fall on a predefined square, or similarly by using a number of random dart throws. This "dartboard algorithm," which can be used in a school setting, shall serve as an example here.

To do this, a chance rain is simulated and the hits counted within and outside the inscribed circle (with radius of length 1) on a square with side length 2. Using dart throws, instead of raindrops, may be a better procedure. With the following considerations, one then reaches a possibility to determine the value of π.

---

16. The name is taken from the gambling paradise.

The relationship between the hits in the circle and the complete number of throws yields an approximation for π:

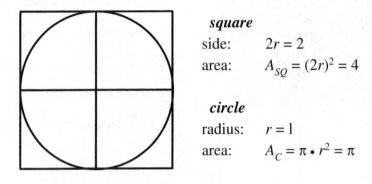

square
side: $2r = 2$
area: $A_{SQ} = (2r)^2 = 4$

circle
radius: $r = 1$
area: $A_C = \pi \cdot r^2 = \pi$

Fig. 3-18

$$\frac{A_C}{A_{SQ}} = \frac{\pi \cdot r^2}{4r^2} = \frac{\pi}{4}$$

$\frac{\pi}{4} =$ probability (circle hits) $= \frac{\text{number of circle hits}}{\text{number of throws}}$

The method yields a good approximation for π only after a very large number of throws. The randomizer (the dart thrower or water dropper) must produce really coincidental numbers and may not be subject to any regularities, that is, the person may not influence where the darts or water droplets fall.

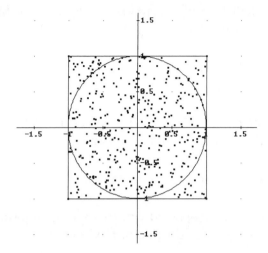

Fig. 3-19

For the first quadrant we get, for example, in the case of ten throws, that the first and the fourth throws don't satisfy the condition $x^2 + y^2 \leq 1$. Therefore, since two points are outside the target region, only eight points are drawn:

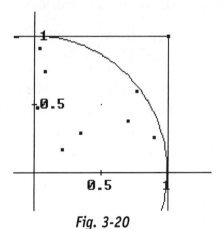

Fig. 3-20

For the calculation of the ratios of the areas, a Monte-Carlo integration is now used. One may proceed as follows:

- With a random numbers generator, an x and a y value between 0 and $2r$ are "thrown."

- The Pythagorean theorem is used to check if the thrown point $P(x, y)$ lies within or outside the circle.

- The hits in the circle are counted.

- The procedure is repeated—the more often repeated the more accurate the expected value of π will be.

Calculating π from a Series of Numbers

Earlier we considered the following formula for π, which was developed by the famous German mathematician Gottfried Wilhelm Leibniz (1646–1716), who, together with Isaac Newton, is credited with developing the modern calculus:

$$\frac{\pi}{4} = 1 - \frac{1}{3} + \frac{1}{5} - \frac{1}{7} + \frac{1}{9} - \frac{1}{11} + \cdots$$

Leibniz, who is also considered one of the great philosophers of the Western world, commented on the unusual connection between the number π and the pattern of alternately adding and subtracting odd unit fractions with the words "Numero deus impare gaudet" (God is happy with the odd number).

We mentioned then that it approaches the value of π rather slowly, since it will take one hundred thousand terms to get to a five-place accuracy of π, and for six-place accuracy we will need to carry this series out for one million terms.

Let's take a look at how this series "behaves." We multiply both sides by 4:

$$\pi = 4 \cdot \left(1 - \frac{1}{3} + \frac{1}{5} - \frac{1}{7} + \frac{1}{9} - \frac{1}{11} + \frac{1}{13} - \frac{1}{15} \pm ...\right) = 4 \cdot \sum_{i=1}^{\infty} \frac{(-1)^{i+1}}{2i-1} \quad \text{[17]}$$

n	Partial Sum	Exact Value	Approximate Value
1	$4 \cdot 1$	4	4
2	$4 \cdot \left(1 - \frac{1}{3}\right)$	$\frac{8}{3}$	2.666666666
3	$4 \cdot \left(1 - \frac{1}{3} + \frac{1}{5}\right)$	$\frac{52}{15}$	3.466666666
4	$4 \cdot \left(1 - \frac{1}{3} + \frac{1}{5} - \frac{1}{7}\right)$	$\frac{304}{105}$	2.895238095
5	$4 \cdot \left(1 - \frac{1}{3} + \frac{1}{5} - \frac{1}{7} + \frac{1}{9}\right)$	$\frac{1,052}{315}$	3.339682539
6	$4 \cdot \left(1 - \frac{1}{3} + \frac{1}{5} - \frac{1}{7} + \frac{1}{9} - \frac{1}{11}\right)$	$\frac{10,312}{3,465}$	2.976046176
7	$4 \cdot \left(1 - \frac{1}{3} + \frac{1}{5} - \frac{1}{7} + \frac{1}{9} - \frac{1}{11} + \frac{1}{13}\right)$	$\frac{147,916}{45,045}$	3.283738483
8	$4 \cdot \left(1 - \frac{1}{3} + \frac{1}{5} - \frac{1}{7} + \frac{1}{9} - \frac{1}{11} + \frac{1}{13} - \frac{1}{15}\right)$	$\frac{135,904}{45,045}$	3.017071817
9	$4 \cdot \left(1 - \frac{1}{3} + \frac{1}{5} - \frac{1}{7} + \frac{1}{9} - \frac{1}{11} + \frac{1}{13} - \frac{1}{15} + \frac{1}{17}\right)$	$\frac{2,490,548}{765,765}$	3.252365934
10	$4 \cdot \left(1 - \frac{1}{3} + \frac{1}{5} - \frac{1}{7} + \frac{1}{9} - \frac{1}{11} + \frac{1}{13} - \frac{1}{15} + \frac{1}{17} - \frac{1}{19}\right)$	$\frac{44,257,352}{14,549,535}$	3.041839618
100	$4 \cdot \left(1 - \frac{1}{3} + \frac{1}{5} - \frac{1}{7} \pm ... - \frac{1}{199}\right)$	numerator and denominator have 88 places (see below)	3.131592903
1,000	$4 \cdot \left(1 - \frac{1}{3} + \frac{1}{5} - \frac{1}{7} \pm ... - \frac{1}{1,999}\right)$	–	3.131592902
10,000	$4 \cdot \left(1 - \frac{1}{3} + \frac{1}{5} - \frac{1}{7} \pm ... - \frac{1}{19,999}\right)$	–	3.141492653
100,000	$4 \cdot \left(1 - \frac{1}{3} + \frac{1}{5} - \frac{1}{7} \pm ... - \frac{1}{199,999}\right)$	–	3.141582653

From the chart above, when $n = 100$ (exact value):

825207975941397038666445468762117443598310111501291263199769614579677862845786070667088

2635106162757236442495826303084698495565581115509040892412867358728390766099042109898375

---

17. This expression is merely a mathematical shorthand way of writing the series.

The approximation value "jumps" back and forth around the actual value of π, since the terms are adding or subtracting alternately; it is, therefore, alternately bigger or smaller than π.

We compare the approximate value for $n = 100,000$ (the last entry in the chart) with the correct value of π = **3.1415**926535897932384..., and we have indeed (only) four correct places. Then as we increase the value of n, the approximation for π becomes increasingly more accurate (i.e., closer to the true value of π).

$n = 1,000,000$:   approximation value = **3.141591**653
$n = 10,000,000$:   approximation value = **3.141592**553
$n = 100,000,000$:   approximation value = **3.141592**643

A Better Series Calculation for the Value of π

There are other series that converge[18] to the value of π faster than the Leibniz series.

Earlier we mentioned the following formula for deriving the value of π, which was discovered by the famous Swiss mathematician Leonhard Euler:

$$\frac{\pi^2}{6} = 1 + \frac{1}{2^2} + \frac{1}{3^2} + \frac{1}{4^2} + \frac{1}{5^2} + \cdots$$

or (after multiplication by a 6 and square root extraction):

$$\pi = \sqrt{6 \cdot \left(1 + \frac{1}{2^2} + \frac{1}{3^2} + \frac{1}{4^2} + \frac{1}{5^2} + \cdots\right)} = \sqrt{6 \cdot \sum_{i=1}^{\infty} \frac{1}{i^2}}$$

This formula is sensationally fast in comparison to that of Leibniz. In the case of the Leibniz series, the computer needed

---

18. To "converge," in this sense, means to approach a particular value as a limit.

more than two and a half hours for $n = 10^8$; the Euler series[19] delivers the accompanying approximation value for $n = 10^8$ in virtually zero seconds of computer time!

The approximate values of π are about the same quality as before. For $n = 10^8$ we get eight-place accuracy: π ≈ **3.141592644**. Let's take a look at some partial sums:

n	Partial Sum	Exact Value	Approximate Value
1	$\sqrt{6 \cdot \frac{1}{1^2}}$	$\sqrt{6}$	2.449489742
2	$\sqrt{6 \cdot \left(1 + \frac{1}{2^2}\right)}$	$\frac{\sqrt{30}}{2}$	2.738612787
3	$\sqrt{6 \cdot \left(1 + \frac{1}{2^2} + \frac{1}{3^2}\right)}$	$\frac{7\sqrt{6}}{6}$	2.857738033
4	$\sqrt{6 \cdot \left(1 + \frac{1}{2^2} + \frac{1}{3^2} + \frac{1}{4^2}\right)}$	$\frac{\sqrt{1,230}}{12}$	2.922612986
5	$\sqrt{6 \cdot \left(1 + \frac{1}{2^2} + \frac{1}{3^2} + \frac{1}{4^2} + \frac{1}{5^2}\right)}$	$\frac{\sqrt{32,214}}{60}$	2.963387701
10	$\sqrt{6 \cdot \left(1 + \frac{1}{2^2} + \frac{1}{3^2} + \cdots + \frac{1}{9^2} + \frac{1}{10^2}\right)}$	$\frac{\sqrt{59,049,870}}{2,520}$	3.049361635
100	$\sqrt{6 \cdot \left(1 + \frac{1}{2^2} + \frac{1}{3^2} + \cdots + \frac{1}{99^2} + \frac{1}{100^2}\right)}$	See Below*	3.132076531
1,000	$\sqrt{6 \cdot \left(1 + \frac{1}{2^2} + \frac{1}{3^2} + \cdots + \frac{1}{999^2} + \frac{1}{1,000^2}\right)}$		3.140638056
10,000	$\sqrt{6 \cdot \left(1 + \frac{1}{2^2} + \frac{1}{3^2} + \cdots + \frac{1}{9,999^2} + \frac{1}{10,000^2}\right)}$		3.141497163
100,000	$\sqrt{6 \cdot \left(1 + \frac{1}{2^2} + \frac{1}{3^2} + \cdots + \frac{1}{99,999^2} + \frac{1}{100,000^2}\right)}$		3.141583104
1,000,000	$\sqrt{6 \cdot \left(1 + \frac{1}{2^2} + \frac{1}{3^2} + \cdots + \frac{1}{\left(10^6\right)^2}\right)}$		3.141591698
10,000,000	$\sqrt{6 \cdot \left(1 + \frac{1}{2^2} + \frac{1}{3^2} + \cdots + \frac{1}{\left(10^7\right)^2}\right)}$		3.141592558
100,000,000	$\sqrt{6 \cdot \left(1 + \frac{1}{2^2} + \frac{1}{3^2} + \cdots + \frac{1}{\left(10^8\right)^2}\right)}$		3.141592644
1,000,000,000	$\sqrt{6 \cdot \left(1 + \frac{1}{2^2} + \frac{1}{3^2} + \cdots + \frac{1}{\left(10^9\right)^2}\right)}$		3.141592652

* $n = 100$ (exact value):

$$\sqrt{476852608239911361933689378555266157910715049563112957966829850441145439327110970306972037522971247716453380893531230355680 0}$$

---

19. Euler also found the sums of the reciprocals of the fourth and sixth powers (by the way, to date no one has been able to do this for third powers!).

For purposes of comparison, here is the value of π:

π = 3.1415926535897932384...

The Genius's Method for Finding the Value of π

The extraordinarily brilliant Indian mathematician Srinivasa Ramanujan (1887–1920) made contributions to the generation of the value of π but left little evidence on how he arrived at his results. Born in 1887 in the small south Indian town of Erode, Ramanujan spent his youth fascinated with mathematics to the detriment of other subjects. As the fledgling Indian Mathematical Society was being founded, it provided a forum for Ramanujan to exhibit his mathematical prowess. For example, in 1911 he posed problems based on his earlier work and found no solvers among the readership. One such example was to evaluate

$$\sqrt{1+2\sqrt{1+3\sqrt{1+4\sqrt{1+\cdots}}}},$$

which appeared harmlessly simple, but yet found no successful solvers. The trick was found in his notebook of theorems that he established. Here, he simply applied the following theorem, which said that if you could represent a number as $(x + n + a)$, the above expression could be represented as

$$x+n+a=\sqrt{ax+(n+a)^2+x\sqrt{a(x+n)+(n+a)^2+(x+n)\sqrt{\cdots}}}$$

So if $3 = x + n + a$, where, say, $x = 2$, $n = 1$, and $a = 0$, then the value of this nest of radicals is simply equal to 3. This is nearly impossible to do without knowledge of Ramanujan's theorem.

With this new exposure, he wrote to three of the top mathematicians in England, E. W. Hobson, H. F. Baker, and G. H. Hardy.[20] Of these three Cambridge professors, only Godfrey Harold Hardy (1877–1947) responded and ultimately invited Ramanujan to England. Hardy thought that the statements contained in the letter had to be correct. For, if they were wrong, nobody would have had such a wild imagination to make them up.

Despite a clash of cultures, the two got along very well and mutually assisted each other. This was the beginning of Ramanujan's popularity outside India. It should be remembered that, even though wearing shoes and using eating utensils were new to this Indian, he came from a long heritage of mathematical culture. The Indians were using our numeration system (including the zero) for over a thousand years before it was introduced in Europe with the publication of Fibonacci's book, *Liber abaci,* in 1202.

---

20. The letter to Hardy, dated "Madras, 16th January 1913" and which enticed Hardy to respond, was the following:

Dear Sir,

I beg to introduce myself to you as a clerk in the Accounts Department of the Port Trust Office at Madras on a salary of only £ 20 per annum. I am now about 23 years of age. I have no University education but I have undergone the ordinary school course. After leaving school I have been employing the spare time at my disposal to work at mathematics. I have not trodden through the conventional regular course which is followed in a University course, but I am striking out a new path for myself. I have made a special investigation of divergent series in general and the results I get are termed by the local mathematicians as "startling.". . .

I would request you to go through the enclosed papers. Being poor, if you are convinced that there is anything of value I would like to have my theorems published. I have not given the actual investigations nor the expressions that I get but I have indicated the lines on which I proceed. Being inexperienced I would very highly value any advice you give me. Requesting to be excused for the trouble I give you.

I remain,

Dear Sir,

Yours truly,

S. Ramanujan.

(As reprinted in Robert Kanigel, *The Man Who Knew Infinity: A Life of the Genius Ramanujan* [New York: Charles Scribner's Sons, 1991], pp. 159–60.)

In our context, Ramanujan came up with some amazing results in the determination of the value of π. He empirically (his word) obtained the approximate value of π with the following expression:

$$\left(9^2 + \frac{19^2}{22}\right)^{\frac{1}{4}} = \left(81 + \frac{361}{22}\right)^{\frac{1}{4}} = \left(\frac{2,143}{22}\right)^{\frac{1}{4}}$$

$$= \left(97.4\overline{09}\right)^{\frac{1}{4}}$$

$$\approx 3.14159265258264612520603717964402237155 7\ldots$$

He further stated that the value he used for π for purposes of calculation was

$$\frac{355}{113}\left(1 - \frac{.0003}{3,533}\right) = 3.1415926535897943\ldots ,$$

which he went on to say "is greater than π by about 10^{-15}" and "is obtained by simply taking the reciprocal of

$$1 - \left(\frac{113\pi}{355}\right) = 11,776,666.61854247437446528035543\ldots .\,"^{[21]}$$

The more ambitious reader can find the justification for this work in appendix B.

Srinivasa Ramanujan gave us other uncanny approximations of π. Even today we are mystified by how he arrived at the various results. Although we are becoming more able to understand his derivations, we still cannot fully appreciate the complexity of the way his unique mind functioned. The following are some of his findings on the value of π.

21. Srinivasa Ramanujan, "Modular Equations and Approximations to π," *Quarterly Journal of Mathematics* 45 (1914): 350–72. Reprinted in *S. Ramanujan: Collected Papers*, ed. G. H. Hardy, P. V. Seshuaigar, and B. M. Wilson (New York: Chelsea, 1962), pp. 22–39.

In chapter 2 we have already mentioned the following formula:

$$\frac{1}{\pi} = \frac{\sqrt{8}}{9,801} \sum_{n=0}^{\infty} \frac{(4n)!(1,103 + 26,390n)}{(n!)^4 396^{4n}}$$

Another formula is

$$\frac{1}{\pi} = \sum_{n=0}^{\infty} \binom{2n}{n}^3 \frac{42n + 5}{2^{12n+4}}$$

The following are some approximations of π that are due to Ramanujan:

$\frac{355}{113} \approx 3.141592920$ was originally discovered by Adriaen Métius (1571-1635),[22] and later Ramanujan gave a geometric construction for this term.

$$\frac{9}{5} + \sqrt{\frac{9}{5}} \approx 3.141640786$$

$$\frac{19\sqrt{7}}{16} \approx 3.141829681$$

Some series discovered by Ramanujan follow. However, the important point is that evaluating such series to huge numbers of digits requires developing specific algorithms.

$$\frac{1}{\pi} = \frac{2\sqrt{2}}{9,801} \sum_{k=0}^{\infty} \frac{(4k)!}{(k!)^4 4^{4k}} \frac{(1,103 + 26,390k)}{99^{4k}} \qquad [\text{Ramanujan}]$$

$$\frac{1}{\pi} = 12 \sum_{k=0}^{\infty} (-1)^k \frac{(6k)!}{(3k)!(k!)^3} \frac{(13,591,409 + 545,140,134k)}{640,320^{3k+3/2}} \qquad [\text{Chudnovsky}]$$

---

22. He and his father, Adriaen Anthoniszoon (c. 1600), took the approximation $3\frac{15}{106} < \pi < 3\frac{17}{120}$, added the numerators (15 + 17 = 32) and the denominators (106 + 120 = 226), took the means (16 and 113), and gave $\pi = 3\frac{16}{113} = \frac{355}{113} = 3.1415929$, a very close approximation. (D. E. Smith, *History of Mathematics*, vol. 2. [New York: Dover, 1958].)

Here are some more of the wondrous discoveries by the genius Ramanujan that you may wish to ponder.

$$\frac{2}{\pi} = 1 - 5\left(\frac{1}{2}\right)^3 + 9\left(\frac{1\cdot3}{2\cdot4}\right)^3 - 13\left(\frac{1\cdot3\cdot5}{2\cdot4\cdot6}\right)^3 + \ldots$$

$$\frac{4}{\pi} = 1 + \left(\frac{1}{2}\right)^2 + \left(\frac{1}{2\cdot4}\right)^2 + \left(\frac{1\cdot3}{2\cdot4\cdot6}\right)^2 + \left(\frac{1\cdot3\cdot5}{2\cdot4\cdot6\cdot8}\right)^2 + \ldots \quad \text{[Forsyth]}$$

$$\frac{1}{\pi} = \frac{1}{72}\sum_{k=0}^{\infty}(-1)^k \frac{(4k)!}{(k!)^4\,4^{4k}} \frac{(23+260k)}{18^{2k}}$$

$$\frac{1}{\pi} = \frac{1}{3{,}528}\sum_{k=0}^{\infty}(-1)^k \frac{(4k)!}{(k!)^4\,4^{4k}} \frac{(1{,}123+21{,}460k)}{882^{2k}}$$

$$\frac{1}{\pi} = 12\sum_{k=0}^{\infty}(-1)^k \frac{(6k)!}{(3k)!(k!)^3} \frac{(A+Bk)}{C^{3k+3/2}} \quad \text{[Borwein]}$$

In the last formula

$$A = 1{,}657{,}145{,}277{,}365 + 212{,}175{,}710{,}912\ \sqrt{61}$$

$$B = 107{,}578{,}229{,}802{,}750 = 13{,}773{,}980{,}892{,}672\ \sqrt{61}$$

$$C = 5{,}280\ (236{,}674 + 30{,}303)\ \sqrt{61}$$

and each additional term in the series adds about thirty-one digits.

We have seen numerous ways that the value of π was calculated. Some were primitive, while others were quite sophisticated. Most remarkably are those that would have appeared to have evolved from spectacular guesses. Today's methods all involve the computer, and how accurate the future calculations of the value of π will be is going to be merely limited by man's creativity and the computer's ability.

Chapter 4

π Enthusiasts

Popularity of π

π is so fascinating and one of the most popular numbers in mathematics for a variety of reasons. First, just understanding what it is (chapter 1) and what it represents and how it can be used has intrigued mathematicians for ages. Its history (chapter 2) over the past four thousand years, spanning the entire globe, has provided amusement and discovery as well as an ongoing challenge. Building upon the continuous attempts at getting ever-more-exact values of π by seeing how many decimal places computers can generate, and how fast they can do it, has become the challenge today for computers and computer scientists, rather than for mathematicians, who still search for more elegant (and efficient) algorithms to accomplish these tasks. Now that we are in the trillions of decimal places, who knows how far we are going to push a computer's capabilities? This seems to be the ultimate test for a computer.

There is a unique curiosity where the enthusiasm for π is demonstrated for all to see. In 1937, in Hall 31 of the Palais de la Decouverte, today a Paris science museum (on Franklin D. Roosevelt Avenue), the value of π was produced with large wooden numerals on the ceiling (a cupola) in the form of a spiral. This was an inspired dedication to this famous number, but there was an error: they used the approximation generated in 1874 by William Shanks, which had an error in the 528th decimal place. This was detected in 1946 and corrected on the museum's ceiling in 1949.

There are many Web sites where π enthusiasts gather to share their latest findings. In the United States, these π lovers celebrate March 14 as **π-day**, since as noted it is 3-14. And at 1:59, they jubilate! (Remember π = 3.14159. . . .) What a coincidence that Albert Einstein was born on March 14, 1879: we can see that this number, 3.141879, is a good approximation of π. Other similar coincidences are constantly found by these π enthusiasts. There are ever more Web sites that help the π-day celebration. Here are just two to begin with: http://www.exploratorium.edu/pi and http://mathwithmrherte.com/pi_day.htm. Here also is the Exploratorium's (San Francisco) announcement of their π-day celebration—celebrating the most famous person born on that day!

Exploratorium welcomes you to:

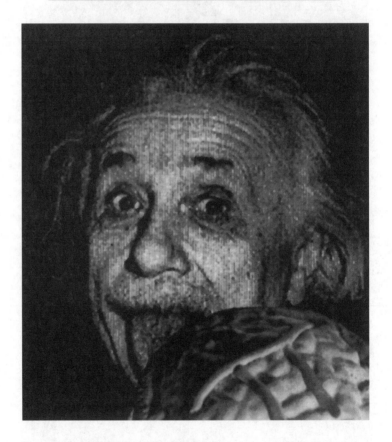

Illustrations copyright © Exploratorium, http://www.exploratorium.edu. Used with permission.

Apparently anything goes when it comes to celebrating π-day. Pose the following question to a mathematician: "What are the next numbers in the following sequence?"

3, 1, 4, 1, 5, . . .

The answer may well be that you have 1s interspersed in the sequence of natural numbers. So the next numbers would be

3, 1, 4, 1, 5, **1, 6, 1, 7, 1, 8, 1, 9,** . . .

However, pose this question to a π enthusiast and the response is surely to be

3, 1, 4, 1, 5, **9, 2, 6, 5, 3, 5, 8, 9, 7, 9, 3, 2, 3, 8, 4, 6, 2, 6,** . . . ,

which, of course, is the value of π (approximately!) This is indicative of the mind-set of π enthusiasts.

The digits of the decimal value of π have been a topic of fascination for centuries. The quest goes on unabated to increase the number of known decimal places of π. Having generated this seemingly endless list of digits comprising the decimal form of π, mathematicians and math enthusiasts have sought ways to find patterns and other entertaining oddities with this number. As with any endless randomly generated list of numbers, you can make just about anything happen within them that you may wish. Here we present a small sample of some of these recreations and oddities.

π Mnemonics

One of the simplest forms of entertainment with this decimal list of digits is to show how many decimal places of the value of π you can commit to memory. Some people like to show off by simply memorizing the first ten, twenty, thirty, or more decimal places. Others who may not have such sharp powers of memory try to create mnemonic devices that will allow them to more easily memorize this list of digits. For your entertainment, we will provide you with a number of these mnemonic devices in a variety of languages, yet from personal experience, a straight memorization of the digits practically lasts forever. Memorize the first twenty-five digits without any device and you will never forget them.

Most of the mnemonic devices for memorizing the decimal value of π require finding somewhat meaningful sentences where the number of letters per word determines the digit.

Although by now most of you have seen the value of π many times, for convenience the first fifty-five decimal places are provided here: **3.14159 26535 89793 23846 26433 83279 50288 41971 69399 37510 58209. . .**

One such sentence used by a number of mathematicians (including Martin Gardner and Howard Eves) is "May I have a large container of coffee?" giving the value 3.1415926, where the *three* letters of "May" give the digit 3, the *one* letter "I" gives the digit 1, the *four* letters of the word "have" give us the digit 4, and so on. A mnemonic that will give us the digits for the first nine decimal places (3.14159265) of π is "But I must a while endeavour[1] to reckon right." We can get digits for the first fourteen decimal places (3.14159265358979) from the sentence "How I want a drink, alcoholic of course, after the heavy lectures involving quantum

---

1. British spelling.

mechanics," which is attributed to James Jeans, Martin Gardner, Howard Eves, and others. A clever mathematician (S. Bottomley) extended this sentence with the phrase "and if the lectures were boring or tiring, then any odd thinking was on quartic equations again," giving us seventeen additional digits and thus the value of π to thirty-one decimal places (3.1415926535897932384626433832795).

People in many countries (and, of course, in a variety of languages) have created poems, jokes, and even dramas where the words used are based on the digits of π. For example, "See, I have a rhyme assisting my feeble brain, its tasks sometime resisting."

We offer here a small collection of such mnemonics, some of which, with the exception of ChiShona and Sindebele, are from the Internet Web site of Antreas P. Hatzipolakis.

Albanian: Kur e shoh e mesoj sigurisht.
[When I see it, I memorize it for sure.] (Robert Nesimi)

Bulgarian: Kak e leko i bqrzo iz(ch)islimo pi, kogato znae(sh) kak.
[How easy and quickly was checked pi if you know how.] (Note: 'ch' and 'sh' are single letters in Bulgarian.)

ChiShona (official language of Zimbabwe): Iye 'P' naye 'I' ndivo vadikanwi. 'Pi' achava mwana.
[P and I are lovers. Pi shall be a brainy child.] (Martin Mugochi, mathematics lecturer, University of Zimbabwe)

Dutch: Eva, o lief, o zoete hartedief uw blauwe oogen zyn wreed bedrogen.
[Eve, oh love, oh sweet darling your blue eyes are cruelly deceived.] (This song was being sung in the sixties, and its inventor has slunk into obscurity.)

English: How I wish I could enumerate pi easily, since all these horrible mnemonics prevent recalling any of pi's sequence more simply.

How I want a drink, alcoholic of course, after the heavy chapters involving quantum mechanics. One is, yes, adequate even enough to induce some fun and pleasure for an instant, miserably brief.

French:

Que j'aime à faire apprendre
Un nombre utile aux sages!
Glorieux Archimède, artiste ingénieux,
Toi, de qui Syracuse loue encore le mérite!
[I really like teaching
a number that is useful to wise men!
Glorious Archimedes, ingenious artist,
You, of whom Syracuse still honors the merit!]

Que j'aime à faire apprendre un nombre utile aux sages!
Immortel Archimède, artiste ingénieux
Qui de ton jugement peut priser la valeur?
Pour moi ton problème eut de pareils avantages.
[I really like teaching a number that is useful to wise men!
Glorious Archimedes, ingenious artist,
Who can challenge your judgment?
For me, your problem had the same advantages.] (Published in 1879 in *Nouvelle Correspondence Mathematique* [Brussels] 5, no. 5, p. 449.)

German:

Wie o! dies π
macht ernstlich so vielen viele Müh!

Lernt immerhin, Jünglinge, leichte Verselein,
Wie so zum Beispiel dies dürfte zu merken sein!
[How oh this π
gives so many people so much trouble!
Learn after all, young fellows, easy little verses,
how such, for example, this ought to be memorized!]

Dir, o Held, o alter Philosoph, du Riesen-Genie!
Wie viele Tausende bewundern Geister,
Himmlisch wie du und göttlich!
Noch reiner in Aeonen
Wird das uns strahlen
Wie im lichten Morgenrot!
[You, oh hero, oh old philosopher, you great genius!
How many thousands admire spirits,
Heavenly as you and godly,
Still more pure in Aeonon
Will beem on us
As in a light dawn.]

Greek:
Αει ο Θεος ο Μεγας γεωμετρει
Το κυκλου μηκος ινα οριση διαμετρω
Παρηγαγεν αριθμον απεραντον
και ον φευ ουδεποτε ολον
θνητοι θα ευρωσι.
[The great God who always works with geometry,
in order to determine the ratio of the circumference of a circle
    to its diameter,
created an infinite number,
that will never be determined in its entirety by mere mortals.]

Italian:

Che n'ebbe d'utile Archimede da ustori vetri sua somma scoperta?
[What advantage did Archimedes' discovery of the burning mirror have?] (Isidoro Ferrante)

Polish:

Kto v mgle i slote
vagarovac ma ochote,
chyba ten ktory
ogniscie zakochany,
odziany vytwornie,
gna do nog bogdanki
pasc kornie.
[Who likes to skip school on a rainy and misty day, perhaps the one who madly in love, smartly dressed, runs to fall humbly at the feet of his loved one.]

Portuguese: Sou o medo e temor constante do menino vadio.
 [I am the constant fear and terror of lazy boys.]

Romanian: Asa e bine a scrie renumitul si utilul numar.
[That's the way to write the famous and useful number.]

Sindebele (official language of Zimbabwe): Nxa u fika e khaya uzojabula na y'nkosi ujesu qobo.
[When you get to heaven, you will rejoice with the Lord Jesus]. (Note: again, the last digit represented here is due to rounding off— it should be 3. Dr. Precious Sibanda, University of Zimbabwe, mathematics lecturer)

Spanish: Sol y Luna y Cielo proclaman al Divino Autor del
Cosmo.
[Sun and Moon and Skies proclaim the divine author of the
Universe.]

Soy π lema y razón ingeniosa
De hombre sabio que serie preciosa
Valorando enunció magistral
Con mi ley singular bien medido
El grande orbe por fin reducido
Fue al sistema ordinario cabal.
[I am pi motto and ingenious reason
of wise man that beautiful series
valuing I enunciate magisterial
with my singular law measured well
the big world finally limited
it went to the ordinary complete system.]
(Columbian poet R. Nieto Paris, according to V. E. Caro, *Los
Numeros* [Bogota: Editorial Minerva, 1937], p. 159.)

Swedish:

Ack, o fasa, π numer fœrringas
ty skolan låter var adept itvingas
räknelära medelst räknedosa
och sa ges tilltron till tabell en dyster kosa.
Nej, låt istället dem nu tokpoem bibringas!
[Oh no, Pi is nowadays belittled
for the school makes each student learn
arithmetic with the help of calculators
and thus the tables have a sad future.
No, let us instead read silly poems!] (Frank Wikström)

For those of you who wish to create a π mnemonic, we offer (for convenience) the value of π to enough places to satisfy most. Remember, there is a limit to how many words one can memorize, even if they produce interesting content. You might be interested to know that the world record holder for the greatest number of digits of π memorized is Hiroyuki Goto, who took over nine hours to recite more than forty-two thousand digits of π.[2]

π = 3.14159 26535 89793 23846 26433 83279 50288 41971 69399 37510 58209 74944 59230 78164 06286 20899 86280 34825 34211 70679 82148 08651 32823 06647 09384 46095 50582 23172 53594 08128 48111 74502 84102 70193 85211 05559 64462 29489 54930 38196 44288 10975 66593 34461 28475 64823 37867 83165 27120 19091 45648 56692 34603 48610 45432 66482 13393 60726 02491 41273 72458 70066 06315 58817 48815 20920 96282 92540 91715 36436 78925 90360 01133 05305 48820 46652 13841 46951 94151 16094 33057 27036 57595 91953 09218 61173 81932 61179 31051 18548 07446 23799 62749 56735 18857 52724 89122 79381 83011 94912 98336 73362 44065 66430 86021 39494 63952 24737 19070 21798 60943 70277 05392 17176 29317 67523 84674 81846 76694 05132 00056 81271 45263 56082 77857 71342 75778 96091 73637 17872 14684 40901 22495 34301 46549 58537 10507 92279 68925 89235 42019 95611 21290 21960 86403 44181 59813 62977 47713 09960 51870 72113 49999 99837 29780 49951 05973 17328 16096 31859 50244 59455 34690 83026 42522 30825 33446 85035 26193 11881 71010 00313 78387 52886 58753 32083 81420 61717 76691 47303 59825 34904 28755 46873 11595 62863 88235 37875 93751 95778 18577 80532 17122 68066 13001 92787 66111 95909 21642 01989

---

2. "Japanese Student Recites Pi to 42,194 Decimal Places," *Seattle Times*, February 26, 1995.

More Fascination with the Digits of π

Then there are those who are fixated on the frequency of the digits of the decimal expansion of π. That is, do the digits come up with equal frequency throughout the many decimal places of π? To determine this, we need to look at a frequency distribution—a table that summarizes the frequency that each of the digits appears within certain intervals. For the first one hundred decimal places of π, do the digits appear with equal frequency? If not, then almost equal frequency? To expect equal frequency within the first one hundred digits would be a bit unreasonable. When we inspect these digits, we discover how far off the distribution is from this exactly equal frequency. There are statistical tests to determine if the slight bit that they may be off for equality is due to chance. If the disparity is due to chance, then we say that the distribution is statistically significantly the same as an equal distribution. This, you will find, is the case in the distribution of the digits yielding the decimal places for the value of π. The following distribution of decimal digits d is found for the first 10^n digits of $π - 3$, that is, we are concerned with only the decimal part of π.[3] It shows no statistically significant departure from a uniform distribution. Dr. Yasumasa Kanada provides the distribution of the first 1.24 trillion places, the world record for the value of π found at the end of 2002.

The number of times the digits appear within the first 10^n places of π

Digits	1 to 10^2	1 to 10^3	1 to 10^4	1 to 10^5	1 to 10^6	1 to 10^7	1 to 10^8	1 to 10^9	1 to 10^{10}	1 to 10^{11}	1 to 10^{12}
0	8	93	968	9999	99959	999440	9999922	99993942	999967995	10000104750	99999485134
1	8	116	1026	10137	99758	999333	10002,475	99997334	1000037790	9999937631	99999945664
2	12	103	1021	9908	100026	1000306	10001092	100002410	1000017271	10000026432	100000480057
3	11	102	974	10025	100229	999964	9998442	99986911	999976483	9999912396	99999787805
4	10	93	1012	9971	100230	1001093	10003863	100011958	999937688	10000032702	100000357857
5	8	97	1046	10026	100359	1000466	9993478	99998885	1000007928	9999963661	99999671008
6	9	94	1021	10029	99548	999337	9999417	100010387	999985731	9999824088	99999807503
7	8	95	970	10025	99800	1000207	9999610	99996061	1000041330	10000084530	99999818723
8	12	101	948	9978	99985	999814	10002180	100001839	999991772	10000157175	100000791469
9	14	106	1014	9902	100106	1000040	9999521	100000273	1000036012	9999956635	99999854780

3. Y. Kanada, "Sample Digits for Decimal Digits of Pi," January 18, 2003, http://www.super-computing.org/pi-decimal_current.html.

Another, and more detailed, distribution is provide by Dr. Kanada. Here you can see that as the number of digits considered increases, the digits come closer to an equal frequency for all digits. In the first one hundred places, there are many more 9s (fourteen) than there are 0s, 1s, 5s, or 7s. Among the first two hundred places, there is less than half the number of 7s as 8s. And so it goes until we get to a larger number of decimal places.

Frequency distribution for the digits of π up to 1,200,000,000,000 decimal places

number of digits considered	0	1	2	3	4	5	6	7	8	9	CHI-SQ.
100	8	8	12	11	10	8	9	8	12	14	4.20
200	19	20	24	19	22	20	16	12	25	23	6.80
500	45	59	54	50	53	50	48	36	53	52	6.88
800	74	92	83	79	80	73	77	75	76	91	5.13
1000	93	116	103	102	93	97	94	95	101	106	4.74
2000	182	212	207	188	195	205	200	197	202	212	4.34
5000	466	532	496	459	508	525	513	488	492	521	10.77
8000	754	833	811	781	809	834	816	786	764	812	8.52
10000	968	1026	1021	974	1012	1046	1021	970	948	1014	9.32
20000	1954	1997	1986	1986	2043	2082	2017	1953	1962	2020	7.72
50000	5033	5055	4867	4947	5011	5052	5018	4977	5030	5010	5.86
80000	7972	8141	7920	7975	7957	8044	8026	8031	7953	7981	4.46
100000	9999	10137	9908	10025	9971	10026	10029	10025	9978	9902	4.09
200000	20104	20063	19892	20010	19874	20199	19898	20163	19956	19841	7.31
500000	49915	49984	49753	50000	50357	50235	49824	50230	49911	49791	7.73
800000	79949	79851	79872	79962	80447	80298	79650	79884	80167	79920	6.27
1000000	99959	99758	100026	100229	100230	100359	99548	99800	99985	100106	5.51
2000000	199792	199535	200077	200141	200083	200521	199403	200310	199447	200691	9.00
5000000	499620	499898	499508	499933	500544	500025	498758	500880	499880	500954	7.88
8000000	799111	800110	799788	800234	800202	800154	798885	800560	800638	800318	3.79
10000000	999440	999333	1000306	999964	1001093	1000466	999337	1000207	999814	1000040	2.78
20000000	2001162	1999832	2001409	1999343	2001106	2000125	1999269	1998404	1999720	1999630	4.17
50000000	4999632	5002220	5000573	4998630	5004009	4999797	4998017	4998895	4998494	4999733	6.17
80000000	7998807	8002788	8001828	7997656	8003525	7996500	7998165	7999389	8000308	8001034	5.95
100000000	9999922	10002475	10001092	9998442	10003863	9993478	9999417	9999610	10002180	9999521	7.27
200000000	19997437	20003774	20002185	20001410	19999846	19993031	19999161	20000287	20002307	20000562	4.13
500000000	49995279	50000437	50011436	49992409	50005121	49990678	49998820	50000320	50006632	49998868	7.42
800000000	79991897	79997003	80003316	79989651	80016073	79961220	80004148	79995109	80002933	80003750	6.62
1000000000	99993942	99997334	100002410	99986911	100011958	99998885	100010387	99996061	100001844	100000273	4.92
2000000000	199994317	199995284	199992575	199999470	200011368	199989852	200004785	199979293	200017844	200012212	6.69
5000000000	499989001	500034127	499984949	499990521	499978284	499995352	500019818	500001703	499990532	500015923	5.68
8000000000	799995840	800031172	800016834	799985886	799942991	799995302	800003383	800012745	800011229	800040618	10.36
10000000000	999967995	1000037790	1000017271	999985483	999937688	1000007928	999985731	1000041330	999917629	1000036012	10.53
20000000000	2000015287	2000017271	2000016834	1999918306	1999950273	2000036170	1999981153	2000034984	2000012969	2000033456	6.88
50000000000	5000012647	5000015776	5000020237	4999914405	5000023598	4999991499	4999928368	5000014860	5000117637	4999990486	5.60
80000000000	8000015632	7999999508	7999996393	7999896897	8000100154	7999913060	7999829394	8000053309	8000156857	7999942311	11.15
100000000000	10000104750	10000026432	10000136978	9999912396	10000032702	9999963661	9999824088	10000084530	10000157175	9999956635	9.03
200000000000	20000030841	20000136978	20000089054	19999918306	20000105099	19999917053	19999881515	19999967594	20000291044	19999669180	8.09
300000000000	29999944911	30000124949	30000089054	29999970781	30000105099	29999898037	29999948243	29999881566	30000345894	29999714598	8.95
400000000000	40000014210	40000141945	40000124949	39999967231	40000267249	39999843336	39999850687	39999950687	40000234425	39999690455	7.44
500000000000	50000008881	50000180765	50000180765	49999950781	50000031060	49999990532	49999986492	49999913528	50000276183	49999708312	5.07
600000000000	59999788154	60000334158	60000439726	59999869412	60000131060	59999819211	59999770829	59999770829	60000439514	59999725725	9.22
700000000000	69999604459	70000439726	70000456638	69999869412	70000162513	69999888676	69999906919	69999845892	70000574684	69999659445	11.61
800000000000	79999579157	80000456158	80000493163	79999787805	80000238690	79999773551	79999935320	79999675065	80000650170	79999802555	12.98
900000000000	89999579157	90000493163	90000480057	89999778899	90000373135	89999836010	89999907911	89999675065	90000761281	89999874525	15.81
1000000000000	99999554071	100000480057	100000527296	99999787805	100000357857	99999807503	99999807503	99999818723	100000791469	99999854788	14.97
1100000000000	110000043750	110000527296	110000620567	109999711592	110000227954	109999687503	109999854514	109999684514	110000827406	109999843711	14.54
1200000000000	119999367735	120000620567	120000620567	119999716885	120000114112	119999710206	119999941333	119999740505	120000830484	119999653604	13.13

Dr. Kanada also provides us with some entertainment within his record-breaking value of π. For example, he spotted the repetition of digits—twelve to be exact—at certain positions of the 1.24 trillion places. Here is a list of these repetitions and the decimal places at which they begin:

777777777777: from 368,299,898,266th decimal place of π
999999999999: from 897,831,316,556th decimal place of π
111111111111: from 1,041,032,609,981th decimal place of π
888888888888: from 1,141,385,905,180th decimal place of π
666666666666: from 1,221,587,715,177th decimal place of π

We also find the sequence of the natural numbers (with zeros at both ends of the sequence) at various places among the first 1.24 trillion places. Here they are along with the place at which they begin:

01234567890 : from 53,217,681,704th decimal place of π
01234567890 : from 148,425,641,592th decimal place of π
01234567890 : from 461,766,198,041th decimal place of π
01234567890 : from 542,229,022,495th decimal place of π
01234567890 : from 674,836,914,243th decimal place of π
01234567890 : from 731,903,047,549th decimal place of π
01234567890 : from 751,931,754,993th decimal place of π
01234567890 : from 884,326,441,338th decimal place of π
01234567890 : from 1,073,216,766,668th decimal place of π

They can also be found in reverse order:

09876543210 : from 42,321,758,803th decimal place of π
09876543210 : from 57,402,068,394th decimal place of π
09876543210 : from 83,358,197,954th decimal place of π
09876543210 : from 264,556,921,332th decimal place of π

09876543210 : from 437,898,859,384th decimal place of π
09876543210 : from 454,479,252,941th decimal place of π
09876543210 : from 614,717,584,937th decimal place of π
09876543210 : from 704,023,668,380th decimal place of π
09876543210 : from 718,507,192,392th decimal place of π
09876543210 : from 790,092,685,538th decimal place of π
09876543210 : from 818,935,607,491th decimal place of π
09876543210 : from 907,466,125,920th decimal place of π
09876543210 : from 963,868,617,364th decimal place of π
09876543210 : from 965,172,356,422th decimal place of π
09876543210 : from 1,097,578,063,492th decimal place of π

These are just a few of the "entertaining" aspects of the decimal value of π. Actually, since the decimal extension will go on indefinitely (even though we now have it only to 1.24 trillion places), one should be able to find any combination of numbers among this sequence of digits. For example, the birthday of the United States (7-4-1776), that is, 741776, appears beginning with the 21,134th decimal place of π. The authors' respective birthdays were found among the first 100 million decimal places of π as follows:

October 18, 1942, written as 10181942, was found beginning at the 1,223rd place of π, and
December 4, 1946, written as 12041946, was found beginning at the 21,853,937th place of π.

You can have fun trying to locate other strings of numbers. The easiest way to do this is to search the Internet for a Web site that does this for you. There are many such available. All you need to do is type the string of numerals you seek to find, and the search engine will find the location of these within a few seconds.

If you take the first string of numerals—314159—to see when it next appears, the search engines will likely tell you that it reappears at the 176,451st place and then reappears another seven times in the first 10 million places of π. So now the rest is for your recreation. Search for your personal string of numbers on any of these search engines. You might begin with your birth date. Remember, if you minimize the number of digits in your birth date, you will have a greater chance of finding it among the known digits of π. So you are better off when searching for April 18, 1944, by searching for 41844, than if you search for 04181944. Some of you might have luck with the longer version as well.

An Optical Illusion

π enthusiasts also focus on the purely geometric stage. Without π they wouldn't be able to discern the following optical illusion, namely, that both inner circles are the same size.

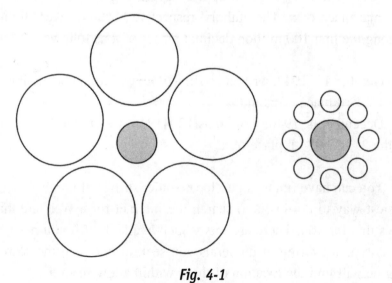

Fig. 4-1

A π Song

As a parting note on the π enthusiasts, we offer the following: a π song! This song, adapted from Don McLean's "American Pie" by Lawrence (Larry) M. Lesser from Armstrong Atlantic State University, gives historical highlights of the number π.

Visit http://www.real.armstrong.edu/video/excerpt1.html to download a video of Larry performing this and some other math songs. We also recommend Larry's "math and music" page at http://www.math.armstrong.edu/faculty/lesser/Mathemusician.html

"AMERICAN π" by Lawrence Lesser (reprinted with permission)

CHORUS: Find, find the value of pi, starts 3 point 14159.
Good ol' boys gave it a try, but the decimal never dies,
The decimal never dies . . .

In the Hebrew Bible we do see
the circle ratio appears as three.
And the Rhind Papyrus does report four-thirds to the fourth,
& 22 sevenths Archimedes found
with polygons was a good upper bound.
The Chinese got it really keen:
three-five-five over one thirteen!
More joined the action
with arctan series and continued fractions.
In the seventeen-hundreds, my oh my,
the English coined the symbol π.
Then Lambert showed it was a lie
to look for rational π.
He started singing . . . (Repeat Chorus)

Late eighteen-hundreds, Lindemann shared
why a circle can't be squared.
But there's no tellin' some people—
can't pop their bubble with Buffon's needle,
Like the country doctor who sought renown
from a new "truth" he thought he found.
The Indiana Senate floor
read his bill that made π four.
That bill got through the House
with a vote unanimous!
But in the end the statesmen sighed,
"It's not for us to decide."
So the bill was left to die
like the quest for rational π.
They started singing . . . (Repeat Chorus)

That doctor's π in the sky dreams
may not look so extreme
If you take a look back: math'maticians long thought that
Deductive systems could be complete
and there was one true geometry.
Now in these computer times,
we test the best machines to find
π to a trillion places
that so far lack pattern's traces.
It's great when we can truly see
math as human history—
That adds curiosity . . . easy as π!
Let's all try singing . . . (Repeat Chorus)

Chapter 5

π Curiosities

The number π has a tendency to pop up when you might least expect it, as was the case with Buffon's needle, where the probability of a tossed needle landing on the lines of a ruled piece of paper led us to a very close approximation of the value of π.

π Digit Curiosities

There are some rather surprising curiosities surrounding the value of π. You might find them coincidental or mysterious. We will let you judge. For example, the circle has 360 degrees, and that fact is connected with π in a peculiar way. Look at the 360th decimal position of π (the 3 before the decimal point is counted):

3.14159265358979323846264338327950288419716939937510582097494459230781640628620899862803482534211706798214808651328230664709384460955058223172535940812848111745028410270193852110555964462294895493038196442881097566593344612847564823378678316527120190914564856692346034861045432664821339360726024914127372458700660631558817488152092096282925409171536436789259031590 **360**

The number **3** is at the 359th place, the number **6** at the 360th place, and the number **0** at the 361st place. This places 360 centered at the 360th digit.

Again, considering the value of π (below) we recall that two of the more accurate fractional approximations of π are

$$\frac{22}{7} = 3.142857142857\overline{142857} \text{ and}$$

$$\frac{355}{113} \approx 3.14159292035398230088495572124$$

We can see that when we locate the 7th, 22nd, 113th, and 355th positions in the decimal value of π, they all have a "2" in that position. Is this coincidental, or does it have some mysterious meaning?

3.14159265358979323846264338327950288419716939937510582097494459230781640628620899862803482534211706798214808651328230664709384460955058223172535940812848111745028410270193852110555964462294895493038196442881097566593344612847564823378678316527120190914564856692346034861045432664821339360726024914127372458700660631558817488152092096282925409171536436789259031590 **360**

This scheme falls apart with the next approximation of π, namely, $\frac{52,163}{16,604} \approx 3.14159238737653577451216574311944$, since, although the 52,163rd place is a "2," the 16,604th place is a "1," although it is preceded and succeeded by a "2." If anything, this prevents us from making a rule about the digits of π, which would not have been true.

Probability's Use of π

It is quite curious that π is related to probability. For example, the probability that a number chosen at random from the set of natural numbers[1] has no repeated prime divisors[2] is $\frac{6}{\pi^2}$. This value also represents the probability that two natural numbers selected at random will be relatively prime.[3] This is quite astonishing since π is derived from a geometric setting.

Using π to Measure the Lengths of Rivers

Another such curious appearance of π arises when we inspect the path of a river. Hans-Henrik Stølum, a geologist at Cambridge University, calculated the ratio between twice the total length of a river and the direct distance between the source and the end of a river.[4] Recognizing that the ratio may vary from river to river, he found the average ratio to be a bit greater than 3. It may be about 3.14, which we recognize as an approximation for π.

Rivers have a tendency to wind back and forth. This so-called meandering of a river is particularly interesting. The term "meandering" came from the river *Maeander*, which is today called Büyük

---

1. The natural numbers are simply our counting numbers: 1, 2, 3, 4, 5, 6, 7, 8, 9, 10, 11, 12,....
2. That means in the set of prime divisors, no prime number will appear more than once.
3. Two numbers are relatively prime when they do not have a common divisor, other than 1.
4. H.-H. Stølum, "River Meandering as a Self-Organizing Process," *Science* 271 (1996): 1710–13.

Menderes (in western Turkey), and it flows in the Aegean Sea at the ancient Milet. This river shows particularly strong meanders.

Albert Einstein was the first to point out that rivers have a tendency toward a loopy path, that is, a slight bend will lead to faster currents on the outside shores, and the river will begin to erode and create a curved path. The sharper the bend, the more strongly the water flows to the outside, and in the consequence the erosion is in turn the faster.

The meanders get increasingly more circular, and the river turns round and returns. It then runs straight ahead again, and the meander becomes a bleak branch. Between the two reverse processes a balance adapts.

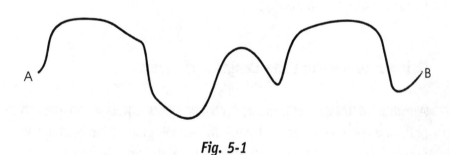

Fig. 5-1

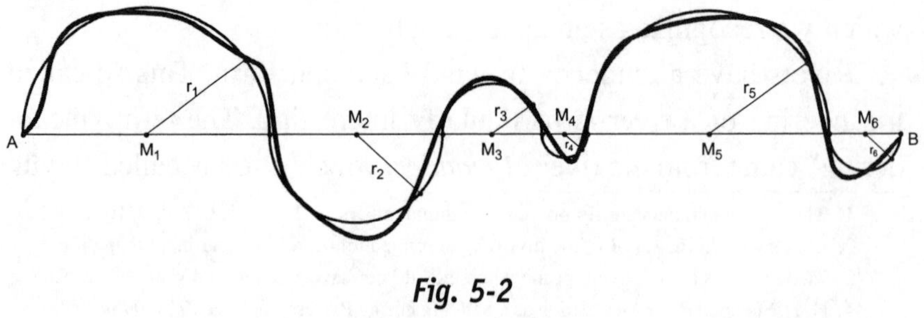

Fig. 5-2

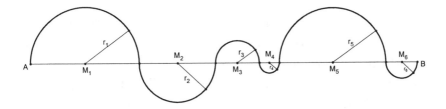

Fig. 5-3

Let us take a look at a fictitious river and superimpose semicircles over the curves. We then have a sum of semicircular arcs that will be compared to a single semicircular arc with a diameter equal to the full distance the river will have traveled (as the crow flies).

l = length of the river from the source A to the mouth B

AB = (straight) distance between the source A and the mouth B

M_i = midpoint of the diameter of the semicircle[5] with radius r_i

a = approximation of the river's length (sum of the semicircles' arcs):

$$a = \pi r_1 + \pi r_2 + \pi r_3 + \pi r_4 + \pi r_5 + \pi r_6 = \pi(r_1 + r_2 + r_3 + r_4 + r_5 + r_6)$$

$$2a = 2\pi(r_1 + r_2 + r_3 + r_4 + r_5 + r_6) = \pi \cdot AB, \text{ which means}$$

$$\frac{\pi}{2} = \frac{a}{AB} \approx \frac{l}{AB} \qquad \text{or} \qquad \pi = \frac{2a}{AB} \approx \frac{2l}{AB}.$$

Rivers that run with a gradual drop in elevation, as can be found in Brazil or in the Siberian tundra, deliver the best approximation for π. From a peculiar application of π, we will now focus our attention on the unusual ways we can express approximations of π.

---

5. This means that the semicircle with diameter midpoint M_i has a radius length r_i.

Unexpected π Coincidences

The value of π comes up in the oddest places. In some cases it "almost appears." Mathematicians for centuries in their quest for establishing the value for π have "collected" these close approximations for π. We offer a small list of some of these very close approximations to π, for example, $\sqrt{10} \approx 3.162278$ is surprisingly close to π. We can continue to list more of these curiously close approximations to π. In some cases, such as with $\sqrt[3]{31} \approx 3.141380652391$, they probably came up by chance and were immediately recognized by the mathematics community (and then, of course, treasured). In other cases, the finding can be considered to border on ingenious—or just lucky? You decide.

Here are a few other "estimates" of π. After inspecting the following list, perhaps you can devise another such approximation of π.

$$\sqrt{2} + \sqrt{3} \approx 3.14626436994$$

$$\frac{333}{106} = 3.14\overline{1509433962264}\,^{6}$$

$$(1.1) \times (1.2) \times (1.4) \times (1.7) = 3.1416$$

$$1.09999901 \times 1.19999911 \times 1.39999931 \times 1.69999961$$
$$\approx 3.141592573$$

$$\left(\frac{355}{113}\right)\left(1 - \frac{0.0003}{3,533}\right) \approx 3.1415926535897943$$

$$\frac{47^3 + 20^3}{30^3} - 1 \approx 3.141592593*$$

---

6. The bar above the digits indicates that those digits repeat in this order indefinitely.

*The four equations marked with an asertisk are from Dario Castellanos, "The Ubiquitous π," *Mathematics Magazine* 61, no. 2 (April 1988).

$$\left(97 + \frac{9}{22}\right)^{\frac{1}{4}} \approx \mathbf{3.14159265}2582646125206037179644$$

$$\left(\frac{77{,}729}{254}\right)^{\frac{1}{5}} \approx \mathbf{3.1415926541}$$

$$\left(31 + \frac{62^2 + 14}{28^4}\right)^{\frac{1}{3}} \approx \mathbf{3.14159265363}*$$

$$\frac{1{,}700^3 + 82^3 - 10^3 - 9^3 - 6^3 - 3^3}{69^3} \approx \mathbf{3.1415926535881}*$$

$$\left(100 - \frac{2{,}125^3 + 214^3 + 30^3 + 37^2}{82^5}\right)^{\frac{1}{4}} \approx \mathbf{3.141592653589780}*$$

$$\frac{9}{5} + \sqrt{\frac{9}{5}} \approx \mathbf{3.1416407864998738}$$

$$\frac{19\sqrt{7}}{16} \approx \mathbf{3.1418296818892}$$

$$\left(\frac{296}{167}\right)^2 \approx \mathbf{3.14159704543}$$

$$2 + \sqrt{1 + \left(\frac{413}{750}\right)^2} \approx \mathbf{3.141592}920$$

$$\left(\frac{63}{25}\right)\left(\frac{17 + 15\sqrt{5}}{7 + 15\sqrt{5}}\right) \approx \mathbf{3.14159265380}$$

$$\sqrt{9.87} = \mathbf{3.141}655614...$$

$$\sqrt{9.8696} = \mathbf{3.14159}1...$$

$$\sqrt{9.869604401} = \mathbf{3.14159265357}...$$

$$\sqrt{9.8696044010893586188344491} =$$
$$\mathbf{3.14159265358979323846264338329}... \qquad \text{[R. Möhwald]}$$

$$\sqrt[4]{9^2 + \frac{19^2}{22}} = \mathbf{3.141592652}... \qquad \text{[Ramanujan]}$$

$$2 + \sqrt[4]{4!} = \mathbf{3.1415}86440...$$

$$\sqrt[4]{\frac{2{,}143}{22}} = \textbf{3.141592652}...$$

$$\sqrt[3]{31 + \frac{25}{3{,}983}} = \textbf{3.1415926534}...$$

$$\sqrt[3]{31} = \textbf{3.14 13 80 6}...$$

$$\left(\sqrt{\sqrt{\sqrt{\sqrt{\sqrt{\sqrt{\sqrt{\sqrt{\sqrt{7}}}}}}}}}\right)^{\sqrt{9!}} = \textbf{3.141603591}...$$

$$\left(\sqrt{\sqrt{\sqrt{\sqrt{\sqrt{\sqrt{\sqrt{\sqrt{\sqrt{7}}}}}}}}}\right)^{\sqrt{9! - \sqrt{\sqrt{4!}}}} = \textbf{3.141592624}...$$

By the way, just for fun, look at this: $\sqrt{\pi} = \textbf{1.772453851}...$ and $\frac{553}{312} = \textbf{1.772}\overline{\textbf{435897}}$, which implies that another good approximation would be $\left(\frac{553}{312}\right)^2 = \textbf{3.14152901}....$

In mathematics we are always looking for connections between concepts that on the surface have nothing to do with each other. For example, a connection between π and the golden ratio,[7] φ, is not easy to find. Yet, Clifford A. Pickover in his book *The Loom of God: Mathematical Tapestries at the Edge of Time* has almost made the connection. He makes the "almost connection" with the following: $\frac{6}{5}$ φ = π. But this is, again, only an approximation of π, since $\frac{6}{5}$φ = 3.1416407864998738178..., while π = 3.1415926535897932384.... So you make the comparison. Satisfied with the connection?

Although not a connection (in the truest sense of the word), another famous mathematical value is the base, *e*, of the natural logarithm, which equals approximately 2.718281828. The value of e^{π} is very close in value to π^{e}. Using a calculator, we can easily calculate each value just to see how close in value these actually are. $e^{\pi} \approx 23.1407...$ and $\pi^{e} \approx 22.4592....$ Quite astonishing![8]

Continued Fractions and π

The value of π can also be expressed as a continued fraction. Before we show this, we will briefly review what a continued fraction is. A continued fraction is a fraction in which the denominator has a mixed number (a whole number and a proper fraction) in it. We can take an improper fraction such as $\frac{13}{7}$ and express it as a mixed number:

$$1\frac{6}{7} = 1 + \frac{6}{7}$$

---

7. The Golden Ratio is the ratio of two line segments, *a* and *b* (where $a < b$), such that $\frac{a}{b} = \frac{b}{a+b}$. The ratio $\phi = \frac{a}{b} = \frac{\sqrt{5}-1}{2} \approx .6180339887498948482045868343656$, while the reciprocal $\frac{b}{a} = \frac{\sqrt{5}+1}{2} \approx 1.6180339887498948482045868343656$. Notice the relationship between the decimals. It suggests that $\phi + 1 = \frac{1}{\phi}$.

8. For the mathematic enthusiast, we provide several proofs of this fact in appendix C. Namely, that $e^{\pi} > \pi^{e}$.

Without changing the value, we could then write this as

$$1 + \frac{6}{7} = 1 + \frac{1}{\frac{7}{6}}$$

which in turn could be written as (again, without any value change)

$$1 + \frac{1}{1+\frac{1}{6}}$$

This is a continued fraction. We would have continued this process, but when we reach a unit fraction, we are essentially finished. Just so that you can get a better grasp of this, we will create another continued fraction. We will convert $\frac{12}{7}$ to a continued fraction form:

$$\frac{12}{7} = 1 + \frac{5}{7} = 1 + \frac{1}{\frac{7}{5}} = 1 + \frac{1}{1+\frac{2}{5}} = 1 + \frac{1}{1+\frac{1}{\frac{5}{2}}} = 1 + \frac{1}{1+\frac{1}{2+\frac{1}{2}}}$$

If we break up a continued fraction into its component parts (called convergents),[9] we get closer and closer to the actual value of the original fraction.

First convergent of $\frac{12}{7} = 1$

Second convergent of $\frac{12}{7} = 1 + \frac{1}{1} = 2$

Third convergent of $\frac{12}{7} = 1 + \frac{1}{1+\frac{1}{2}} = 1 + \frac{2}{3} = 1\frac{2}{3} = \frac{5}{3}$

---

9. This is done by considering the value of each portion of the continued fraction up to each plus sign, successively.

Fourth convergent of $\frac{12}{7} = 1 + \cfrac{1}{1+\cfrac{1}{2+\cfrac{1}{2}}} = \frac{12}{7}$

The above examples are all finite continued fractions. They result in rational numbers (those that can be expressed as simple fractions—albeit improper fractions). It would then follow that an irrational number would result in an infinite continued fraction. That is exactly the case. A simple example of an infinite continued fraction is that of $\sqrt{2}$.

$$\sqrt{2} = 1 + \cfrac{1}{2+\cfrac{1}{2+\cfrac{1}{2+\cfrac{1}{2+\cfrac{1}{2+\cfrac{1}{2+\cdots}}}}}}$$

We have a short way to write a long (in this case infinitely long) continued fraction: [1, 2, 2, 2, 2, 2, 2, 2,…], or when there are these endless repetitions, we can even write it in shorter form as $\left[1, \overline{2}\right]$, where the bar over the 2 indicates that the 2 repeats endlessly.

The German mathematician Johann Heinrich Lambert (1728–1777) was the first to rigorously prove that π was irrational. His method of proof was to show that if n is rational (and not zero), then the tangent of n cannot be rational. He said that since $\tan \frac{\pi}{4} = 1$ (a rational number), then $\frac{\pi}{4}$ or π cannot be rational.[10] In 1770 Lambert produced a continued fraction for π.

---

10. Lambert's proof was strengthened by Adrien-Marie Legendre (1752–1833) in his 1794 book, *Éléments de géometrie*.

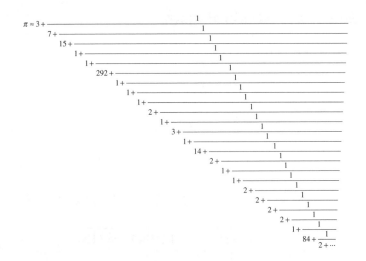

This written in short form is [3, 7, 15, 1, 292, 1, 1, 1, 2, 1, 3, 1, 14, 2, 1, 1, 2, 2, 2, 2, 1, 84, 2,…]

The convergents of this continued fraction are $\frac{3}{1}, \frac{22}{7}, \frac{333}{106}, \frac{355}{113}, \frac{103,993}{33,102}, \frac{104,348}{33,215}, \ldots$

You may remember seeing the first convergents before. They were historical approximations:

3 was the approximation mentioned in the Bible (I Kings 7:23 and 2 Chronicles 4:2).

$\frac{22}{7}$ was the upper bound given by Archimedes in the third century BCE.

$\frac{333}{106}$ was the lower bound for π found by Adriaen Anthoniszoon.

$\frac{355}{113}$ was found about 480 by Tsu Ch'ung Chi and others.

The first four may appear familiar to you since we encountered these approximations earlier. With each successive convergent, we get closer to the value of π. Here are the decimal values of these convergents. Notice how they approach π gradually, each successive one gets ever closer to π.

Convergents of π	Decimal equivalents
$\frac{3}{1}$	3.0
$\frac{22}{7}$	$3.142857\overline{142857}$
$\frac{333}{106}$	$3.141509\overline{433962264}$
$\frac{355}{113}$	$\approx 3.14159292035398230088495575522124$
$\frac{103,993}{33,102}$	$\approx 3.14159265301190260407226149477737$
$\frac{104,348}{33,215}$	$3.1415926539214210\overline{4470871594}$

For the motivated reader, we provide two nonsimple continued fractions (with numerators other than 1) whose successive convergents will also approach the value of π:[11]

$$\frac{4}{\pi} = 1 + \cfrac{1^2}{2 + \cfrac{3^2}{2 + \cfrac{5^2}{2 + \cfrac{7^2}{2 + \cfrac{9^2}{2 + \cfrac{11^2}{2 + \cdots}}}}}}$$

$$\frac{\pi}{2} = 1 + \cfrac{1}{1 + \cfrac{1 \cdot 2}{1 + \cfrac{2 \cdot 3}{1 + \cfrac{3 \cdot 4}{1 + \cfrac{4 \cdot 5}{1 + \cdots}}}}}$$

---

11. In 1869, James Joseph Sylvester (1814–1897) discovered the second continued fraction, shown here. He is also known for his role in founding the *American Journal of Mathematics*.

In peculiar ways we can make π relate to other aspects of mathematics. For example, the harmonic series

$$1 + \frac{1}{2} + \frac{1}{3} + \frac{1}{4} + \frac{1}{5} + \frac{1}{6} + \frac{1}{7} + \frac{1}{8} + \frac{1}{9} + \frac{1}{10} + \ldots$$

in which the addends merely are a sequence formed by taking the reciprocals of the natural numbers: 1, 2, 3, 4, 5, 6, 7, 8, 9, 10, ...

Can this also relate to π? This time, however, we must make a slight modification. We will take the squares of the terms of the harmonic series to get $\frac{\pi^2}{6}$. That is, $\frac{\pi^2}{6} = 1 + \frac{1}{2^2} + \frac{1}{3^2} + \frac{1}{4^2} + \frac{1}{5^2} + \frac{1}{6^2} + \ldots$.

Some other series[12] that relate to π are provided below:

$$\frac{\pi^2}{12} = 1 - \frac{1}{2^2} + \frac{1}{3^2} - \frac{1}{4^2} + \frac{1}{5^2} - \frac{1}{6^2} + \ldots$$

$$\pi = \frac{4}{1} - \frac{4}{3} + \frac{4}{5} - \frac{4}{7} + \frac{4}{9} - \frac{4}{11} + \frac{4}{13} - \frac{4}{15} + \ldots$$

The above expression of π was developed by Leonhard Euler, who also came up with another interesting expression for obtaining the value of π:

$$\frac{2}{\pi} = \left(1 - \frac{1}{4}\right)\left(1 - \frac{1}{16}\right)\left(1 - \frac{1}{36}\right)\left(1 - \frac{1}{64}\right)\left(1 - \frac{1}{100}\right)\ldots,$$

which, by using some elementary algebra,[13] can be written in a simpler form as[14]

$$\frac{2}{\pi} = \left(\frac{1 \cdot 3}{2 \cdot 2}\right)\left(\frac{3 \cdot 5}{4 \cdot 4}\right)\left(\frac{5 \cdot 7}{6 \cdot 6}\right)\left(\frac{7 \cdot 9}{8 \cdot 8}\right)\left(\frac{9 \cdot 11}{10 \cdot 10}\right)\ldots$$

---

12. A series is the sum of the terms of a sequence.

13. The general term can be written as $1 - \frac{1}{n^2}$, which then equals $\frac{n^2 - 1}{n^2} = \frac{(n-1)(n+1)}{n^2}$.

14. This was first developed independently by John Wallis.

While discussing expressions that can represent π, we should note the formula that Leonhard Euler developed:[15]

$$\pi = \lim_{n \to \infty} \left[\frac{1}{n} + \frac{1}{6n^2} + 4n \left(\frac{1}{n^2 + 1^2} + \frac{1}{n^2 + 2^2} + \cdots + \frac{1}{n^2 + n^2} \right) \right]$$

It is interesting to see this formula applied to successive values of n. You will notice that after $n = 10$, the approach to π gets markedly slower.

Values of n	Values of π as determined by Euler's formula
1	3.16666666666666̄
2	3.14166666666666̄
3	3.14159544̄159544
4	3.141593137254902
5	3.141592780477657
10	3.141592655573826
20	3.141592653620795
30	3.141592653592515
50	3.141592653589920
100	3.141592653589795
112	3.141592653589793

We shall end this chapter with some purely recreational illustrations. Dario Castellanos, in his comprehensive article "The Ubiquitous π,"[16] shows how (somewhat circuitously) the number 666 is

---

15. Discovered in a correspondence to Christian Goldbach, Castellanos, "The Ubiquitous π," p. 73.
16. Ibid.

related to π. Be patient as you follow along. First, a word about the number 666. It is the number of the beast in the book of Revelations in the Bible: "Here is wisdom. Let him that hath understanding count the number of the beast; for it is the number of a man, and his number is six hundred, three score and six." It is also the thirty-sixth triangular number $\left(666 = \frac{1}{2} \cdot 36 \cdot 37\right)$. It is also curious that 666 represented in Roman numerals is DCLXVI, which uses all the symbols less than M exactly once.

The number 666 is equal to the sum of the squares of the first seven prime numbers:

$$666 = 2^2 + 3^2 + 5^2 + 7^2 + 11^2 + 13^2 + 17^2$$

Some other peculiarities of 666 follow.

The exponents reflect the number 666 and the bases are the first three natural numbers.

$$666 = 1^6 - 2^6 + 3^6$$

Now look at how the 666 manifests itself:

$$666 = 6 + 6 + 6 + 6^3 + 6^3 + 6^3$$

or $666 = (6 + 6 + 6)^2 + (6 + 6 + 6)^2 + 6 + 6 + 6$

Notice the pattern here:

$$666 = 1^3 + 2^3 + 3^3 + 4^3 + 5^3 + 6^3 + 5^3 + 4^3 + 3^3 + 2^3 + 1^3$$

Having now established the unusualness of the number 666, we will come back to it shortly. Consider the first nine digits of the value of π in groups of three: 314 159 265. The second two groups of three, together with 212, form a Pythagorean triple (159, 212, 265), which means that $159^2 + 212^2 = 265^2$.

Now here is the "stretch." This newly introduced number, 212, together with 666, forms a quotient that gives a nice approximation of π. That is, $\frac{666}{212} = 3.14150943396226$.

A further connection with the relationship of 666 and π: the sum of the first 144 (= [6 + 6] • [6 + 6]) digits of π is 666.

Another recreational application of π Castellanos shows involves a magic square.[17] Consider the conventional 5×5 magic square:

17	24	1	8	15
23	5	7	14	1
4	6	13	20	22
10	12	19	21	3
11	18	25	2	9

We now replace each number with that number digit of the π decimal value. That is, we replace 17 with 2, since 2 is the seventeenth digit in the value of π, and so on.

					Sum of the rows
2	4	3	6	9	24
6	5	2	7	3	23
1	9	9	4	2	25
3	8	8	6	4	29
5	3	3	1	5	17
Sums of the columns 17	29	25	24	23	

Notice how the sums of the columns are the same as the sums of the rows!

You can call the following coincidence or consider it a strange mystery, but look at this next relationship.

---

17. A magic square is a square arrangement of numbers where the sum of each row, column, and diagonal is the same.

Let's look at the first three decimal places of π: 141. The sum of these digits is 6, the first perfect number,[18] and the third triangular number.[19]

Now look at the first seven decimal places of π: 1415926. Their sum is 28, which is the second perfect number and the seventh triangular number. Astonishing symmetry!

Mike Keith's World of Words & Numbers (http://users.aol.com/s6sj7gt/mikehome.htm#toc) provides some unusual numerical recreations. One is an unusual pattern of the digits of π. First, arranging the digits of the decimal value of π as hexagonal numbers,[20] we get the last number (the first hexagonal number) as six nines.

Notice how the last row of digits, representing the first polygonal number, 1, consists of all nines. That is, we ended up at these six nines after the 768th digit.

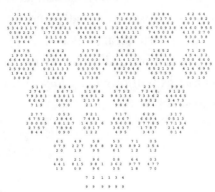

----

18. A perfect number is one where the sum of its proper factors is equal to the number itself. For example, 6 is a perfect number because the sum of its proper factors, 1 + 2 + 3, equals 6.

19. Triangular numbers are those that represent an equilateral array of points:

20. Hexagonal numbers are those that represent a hexagonal array of points:

Hence, 1, 7, 19, etc. are hexagonal numbers.

Realizing that the six nines will appear after the 768th digit, let us now repeat this for 12 × 8 rectangles:

```
314159265358  979323846264  338327950288   419716939937  510582097494  459230781640
628620899862  803482534211  706798214808   651328230664  709384460955  058223172535
940812848111  745028410270  193852110555   964462294895  493038196442  881097566593
344612847564  823378678316  527120190914   564856692346  034861045432  664821339360
726024914127  372458700660  631558817488   152092096282  925409171536  436789259036
001133053054  882046652138  414695194151   160943305727  036575959195  309218611738
193261179310  511854807446  237996274956   735188575272  489122793818  301194912983
367336244065  664308602139  494639522473   719070217986  094370277053  921717629317


              675238  467481  846766   940513  200056  812714
              526356  082778  577134   275778  960917  363717
              872146  844090  122495   343014  654958  537105
              079227  968925  892354   201995  611212  902196


                 086  403  441    815  981  362
                 977  477  130    996  051  870


                    7   2   1    1   3   4
                    9   9   9    9   9   9
```

There are lots of properties that can be established for the number 768. For example, $768 = 3 \times 256 = 3 \times 4^4 = 12 \times 4^3 = (6)(1 + 1 + 2 + 4 + 8 + 16 + 32 + 64)$, as well as others that you can find. These properties allow us to neatly end up with the row of nines in the above geometric arrangements.

Searching the Internet or reading books on number theory and recreational mathematics will provide you with a boundless supply of π peculiarities to savor.

Chapter 6

Applications of π

We will now explore the various applications of π in a variety of ways. This will involve some unusual properties of the circle, which determines π. We will explore the areas of some rather strange-looking regions that are based on the circle and find the lengths of circular arcs that are a bit "off the beaten path." Yet we will begin with the introduction to a geometric shape that shares many properties with the circle but isn't one.

π When You Least Expect It

It is well known that π is related to the circle—as its ratio of circumference to diameter. We begin by inspecting another geometric figure, in which the ratio of its perimeter to its "distance across" is also π. As with the circle, which has a constant breadth, namely, its diameter, this figure also has a constant breadth, although that property is not as obvious as with the circle. The figure of which we speak is very simply constructed. We will introduce it through its construction. We begin by constructing an equilateral triangle and then drawing three congruent circles, using each vertex of the triangle as a center and each radius equal to the side of the triangle.

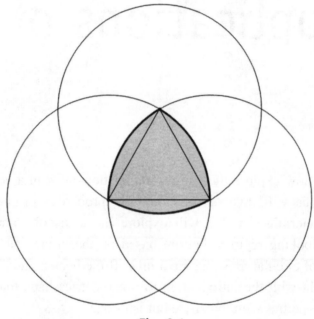

Fig. 6-1

The shaded figure is the subject of this chapter. This shape (seen isolated in fig. 6-2) is called a Reuleaux triangle, named after the German engineer Franz Reuleaux (1829–1905), who taught at the Royal Technical University of Berlin. One might wonder how Franz Reuleaux ever thought of this triangle. It is said that he was in search of a button that was not round but still could fit through a button hole equally well from any orientation. His "triangle" solved the problem, as we will see in the following pages.

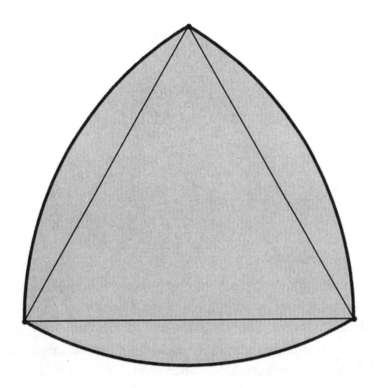

Fig. 6-2

This Reuleaux triangle has many unusual properties. It compares nicely to the circle of similar breadth.[1] What do we mean by the breadth of the Reuleaux triangle? We refer to the distance between two parallel lines tangent to the curve (see fig. 6-3) as the breadth of the curve. Now look carefully at the Reuleaux triangle and notice that no matter where we place these parallel tangents, they will always be the same distance apart—namely, the radius of the arcs comprising the triangle. (See fig. 6-3.)

---

1. In the case of a circle, the breadth is the diameter, while for the Reuleaux triangle, it is the distance across—from a triangle vertex to the opposite arc.

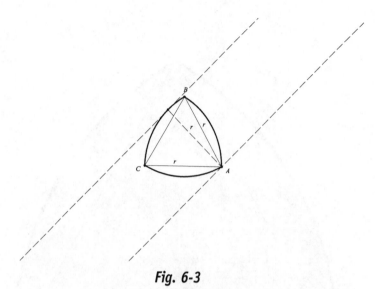

Fig. 6-3

Another geometric figure having a constant breadth is a circle. As you can plainly see in figure 6-4, the "breadth" of a circle is its diameter. The same property holds true for the circle as it did for the Reuleaux triangle: wherever we place the parallel tangents, they will always be the diameter-distance apart.

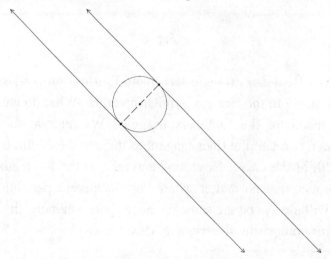

Fig. 6-4

Before we inspect some of the fascinating properties of this Reuleaux triangle, such as the fact that it is analogous to the circle in its ratio of perimeter to breadth also equaling π, we will discuss a "practical application" of the Reuleaux triangle.

You know that if you were to try to turn a circular screw with a normal wrench, you would have no success. The wrench would slip and not allow a proper grip on the circular head of the screw. The same would hold true for a Reuleaux triangular head. It, too, would slip since it is a curve of constant breadth, just like the circle is.

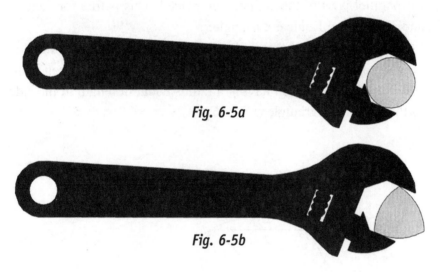

Fig. 6-5a

Fig. 6-5b

So here is a practical application of this situation. During the summer months, kids in a city like to "illegally" turn on the fire hydrants to cool off on very hot days. Since the valve of the hydrant is usually a hexagonal-shaped nut, they simply get a wrench to open the hydrant. If that nut were the shape of a Reuleaux triangle, then the wrench would slip along the curve just as it would along a circle. However, with the Reuleaux triangle nut, unlike a circle-shaped nut, we could have a special wrench with a congruent Reuleaux triangle shape that would fit about the nut and not slip. This would not be possible with a circular nut. Thus, the fire department would be equipped with a special Reuleaux wrench to open the hydrant in cases of fire, yet the Reuleaux triangle could protect against playful water opening and avoid water being wasted in this manner. Just as a matter of curiosity, the fire hydrants in New York City have pentagonal nuts, which also do not have parallel opposite sides and cannot be turned by a normal wrench.

The Reuleaux triangle is said to be, like the circle, a closed curve of constant breadth. That is to say that when one measures the figure with calipers,[2] it will have the same measure no matter where the parallel jaws of the calipers are placed. This is true for a circle and also for the Reuleaux triangle.

As we showed before, the Reuleaux triangle is formed by drawing circles, each centered at a different vertex of a given equilateral triangle and each having a radius equal in length to the side of the equilateral triangle (fig. 6-6).

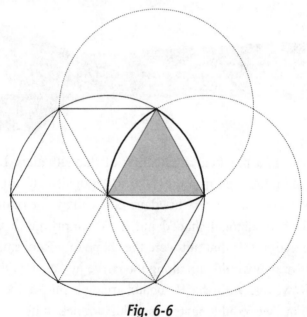

Fig. 6-6

2. An instrument having a fixed and a movable arm on a graduated stock, used for measuring the diameters of logs and similar objects.

Here then is the constructed Reuleaux triangle (fig. 6-7).

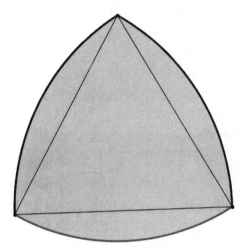

Fig. 6-7

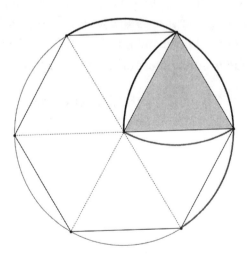

Fig. 6-8

Surprisingly, the circumference of the Reuleaux triangle of breadth r has exactly the same perimeter (i.e., the circumference) as that of a circle with the diameter equal to the breadth of the Reuleaux triangle. We shall verify this relationship between the circle and the Reuleaux triangle.

In figure 6-8 we notice that one "side" of the Reuleaux triangle is one-sixth of the circumcircle of a regular hexagon, so three times this side length give us the perimeter. Therefore, the Reuleaux triangle (of breadth r) has a perimeter that equals

$$3\left[\frac{1}{6}(2\pi r)\right] = \pi r$$

The circle with a diameter of length r has a circumference that is πr, which is the same as the perimeter of the Reuleaux triangle.

The comparison of the areas of these two figures is quite another thing. The areas are not equal. Let's compare the areas.

We can get the area of the Reuleaux triangle in a clever way, by adding the three circle sectors[3] that overlap the equilateral triangle and then deducting the pieces that overlap, so that this region is actually only counted once and not three times. (This will be a very useful technique to remember for use later in this chapter.)

The total area of the three overlapping circle sectors

$$= 3\left(\frac{1}{6}\right)\left(\pi r^2\right)$$

The area of the equilateral triangle[4]

$$= \frac{r^2\sqrt{3}}{4}$$

---

3. A circle sector, which looks like a piece of pie, is a region bounded by two radii of a circle and the circle's arc joining them.

4. This is an important formula to remember and will be used rather frequently. It is obtained by using the Pythagorean theorem to find the altitude, and then simply applying the traditional formula for the area of a triangle: $A = \frac{1}{2}bh$.

The area of the Reuleaux triangle[5] is

$$3\left(\frac{1}{6}\right)\left(\pi r^2\right) - 2\left(\frac{r^2\sqrt{3}}{4}\right) = \frac{r^2}{2}\left(\pi - \sqrt{3}\right)$$

The area of a circle with diameter of length r is

$$\pi\left(\frac{r}{2}\right)^2 = \frac{\pi r^2}{4}$$

Comparing the areas of these two figures of equal breadth indicates that the area of the Reuleaux triangle is less than the area of the circle. This is consistent with our understanding of regular polygons, where the circle has the largest area for a given diameter.

The Austrian mathematician Wilhelm Blaschke (1885–1962) proved that given any number of such figures of equal breadth, the Reuleaux triangle will always possess the smallest area, and the circle will have the greatest area.

Let's now go back and see why the Reuleaux triangle has the same ratio of perimeter to breadth as the circle—namely, π. The perimeter is comprised of three arcs (see fig. 6-8), each one-sixth of a circle of radius, say, r. Therefore the perimeter is

$$3\left[\frac{1}{6}\left(2\pi r\right)\right] = \pi r$$

The breadth is r. So the ratio of perimeter to breadth is $\frac{\pi r}{r} = \pi$, which is exactly what we know about a circle—that the ratio of its perimeter (i.e., circumference) to its breadth (diameter) is equal to π.

We know a wheel rolls on a flat surface quite smoothly. If the Reuleaux triangle is "equivalent" to the circle, it, too, should be able to roll on a flat surface. Well, it can, but it wouldn't be a smooth roll because of the "pointed" corners. Yet if furniture

---

5. We are subtracting two overlapping triangle areas from the three overlapping sectors.

movers would use a roller in the shape of a Reuleaux triangle instead of the usual round, circle-shaped roller, the furniture would not "bounce" the object being moved, but it would roll somewhat irregularly. Notice that the center point (or centroid) of the rolling Reuleaux triangle will not stay at a constant parallel path to the surface being rolled on, as is the case for a circle. The end view of these rolling Reuleaux triangles might look like the following.

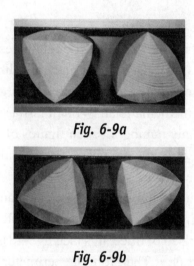

Fig. 6-9a

Fig. 6-9b

We can make an adjustment to the Reuleaux triangle to give it rounded corners, and without destroying its properties.

If we extend the sides (length s) of the equilateral triangle that was used to generate the Reuleaux triangle by an equal amount (say, a) through each vertex, and then draw six circular arcs alternately with the vertices of the triangle as centers (see fig. 6-10), and radius a, the result is a modified Reuleaux triangle with "rounded corners" to allow a smoother roll.

We now need to see that this modified Reuleaux triangle is of constant breadth and that the ratio of its perimeter to its breadth is π.

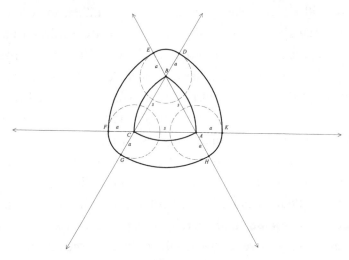

Fig. 6-10

The sum of the lengths of the three smaller "corner arcs" is

$$3\left[\frac{1}{6}(2\pi a)\right]$$

The sum of the lengths of the three larger "side arcs" is

$$3\left[\frac{1}{6}2\pi(s+a)\right]$$

The sum of the six arcs (i.e., the perimeter) is $\pi(s+2a)+\pi a = \pi(s+a)$. The breadth is $(s+2a)$, so the ratio of perimeter to breadth is π. When you would least expect it, π, again, shows up. Comparatively speaking, a circle with diameter $(s+2a)$ has a circumference of $\pi(s+2a)$, the same as the Reuleaux triangle.

Another astonishing property of the Reuleaux triangle is that a drill bit in the shape of a Reuleaux triangle could bore a square hole rather than the expected round hole. Or to put this another way, the Reuleaux triangle is always in contact with each side of a square of appropriate size. This can be seen below (see fig. 6-11). Remember, however, that this drill will not be rotating on a fixed axis; rather, the center of a Reuleaux triangle rotating in the square almost describes a circle— more exactly, it consists of four elliptical arcs. (The circle is the only curve of constant breadth that has a balanced center of symmetry.)

The English engineer Harry James Watt,[6] who lived in Turtle Creek in Pennsylvania, recognized this in 1914, when he received a US patent (no. 1241175), enabling these drills to be produced. The production of drills that can cut square holes was begun in 1916 by the Watt Brothers Tools Factories in Wilmerding, Pennsylvania. Thus, the Reuleaux triangle can be rotated so that it always touches the sides of a square, and thereby brushes over the sides of the square and also gets very close to the corners of the square.

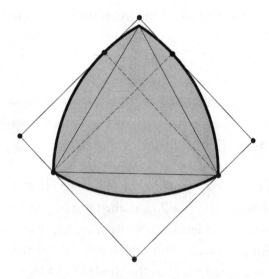

Fig. 6-11

---

6. A descendant of the famous inventor James Watt (1736–1819).

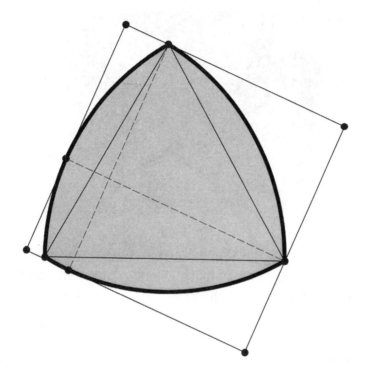

Fig. 6-12

Felix Wankel (1902–1988), a German engineer, built an internal combustion engine for a car that was the shape of a Reuleaux triangle and rotated in a chamber. It had fewer moving parts and gave out more horsepower for its size than the usual piston engines. The Wankel engine was first tried in 1957 and then put into production in the 1964 Mazda. Again, the unusual properties of the Reuleaux triangle made this type of engine possible.

The Energon, in Ulm, Germany, is purported to be the biggest passive office block in the world. It has the outer shape of a Reuleaux triangle and is a low-energy building, heated by geothermal energy.

Fig. 6-13

There are lots of entertaining and useful ideas attached to this Reuleaux triangle, which is the analogue of the circle, and hence shares ownership of π with the circle.

π in Sports

Have your ever wondered how the start positions at a track meet are calculated? Well, this can't be done without π. The standard track is 400 meters, and the width of each runner's lane is 1.25 meters. The track is composed of two straight paths and two semicircular paths.

There are a number of questions that arise in the construction of a racetrack. How long is each lane of the track? How much of a head start, v, in meters should each successive runner have after the runner in lane 1? What must the radius of each of the semicircular parts be in order for lane 1 of the track to be 400 meters long?

We will consider the length of the straight parts of the track to be a meters and the width of each lane to be b meters.

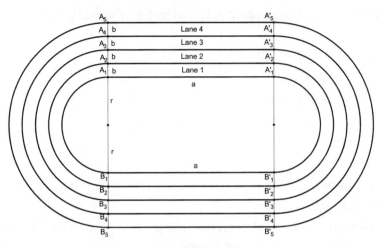

Fig. 6-14

(*Note:* The measurement shall be made 20 cm from the inside edge of each successive lane.)

We begin by measuring lane 1, and then, for each successive lane, we make the proper adjustments as shown below.[7]

lane 1: $C_1 = 2a + 2\pi(r + .2) = 2a + 2\pi r + 2\pi \cdot .2; v_1 = 0$

lane 2: $C_2 = 2a + 2\pi(r + b + .2) = 2a + 2\pi r + 2\pi b + 2\pi \cdot .2;$
$v_2 = 2\pi b$

lane 3: $C_3 = 2a + 2\pi(r + 2b + .2) = 2a + 2\pi r + 4\pi b + 2\pi \cdot .2; v_3$
$= 4\pi b$

lane 4: $C_4 = 2a + 2\pi(r + 3b + .2) = 2a + 2\pi r + 6\pi b + 2\pi \cdot .2; v_4$
$= 6\pi b$

With $a = 100$ m, $b = 1.25$ m, and $C_1 = 2a + 2\pi r + 2\pi \cdot .2$, we get

$2(a + \pi r + .2\pi) = 400$; therefore, $r = \dfrac{100}{\pi} - \dfrac{1}{5} = \dfrac{500 - \pi}{5\pi} \approx 31.63$ m

---

7. 20 cm = .2 m

The handicaps have been calculated in the following way:

$$v_2 = 2\pi b \approx 7.85 \text{ m}$$

$$v_3 = 4\pi b \approx 15.71 \text{ m}$$

$$v_4 = 6\pi b \approx 23.56 \text{ m}$$

Remember, none of this would have been possible without our trusty π!

Not only does π play an ever-important role in finding areas of circles and sections of circles, but now we must use some interesting techniques that will result in perhaps new ways of "looking" at some problem situations—that is, "backing into" the solution, a somewhat indirect method. As we go along from problem to problem, the technique will become more obvious and, we hope, familiar.

A Spiral Formed by Semicircles

We begin by looking at the figures below (fig. 6-15). They appear to be spirals, and can be considered so. However, they are unusual in that they are created by successively larger semicircles. Using the ubiquitous π, we will be able to measure aspects of these spirals: length and area. In figures 6-15a and 6-15b, the points M_u and M_o are at a distance a from each other and are alternately the respective centers of semicircles. M_u is the center of the "bottom" semicircles, and M_o is the center of the semicircles lying above the horizontal diameter.

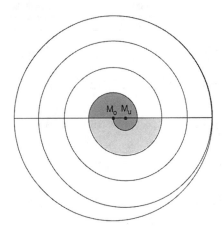

Fig. 6-15a

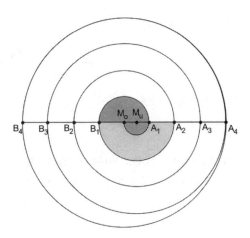

Fig. 6-15b

With the help of our trusty π, we can find the length of the spiral and the areas of the semi-annuli.[8] We will calculate each portion separately. Semicircle c_1 (M_u, a) refers to the semicircle with center M_u and radius length a. The table below shows the calculation for each.

---

8. An annulus is the region between two concentric circles. The semi-annulus is the region between concentric semicircles.

Semicircle	Arc length	Area of the semicircles	Area of the semi-annuli	
$c_1(M_u, a)$	$b_1 = \pi \cdot a$	$A_{sc\,1} = \frac{1}{2}\pi \cdot a^2$	$A_1 = \frac{1}{2}\pi \cdot a^2$	$= \pi\frac{1}{2}a^2$
$c_2(M_0, 2a)$	$b_2 = 2\pi \cdot a$	$A_{sc\,2} = \frac{1}{2}\pi \cdot (2a)^2$	$A_2 = 2\pi \cdot a^2$	$= 2\pi a^2$
$c_3(M_u, 3a)$	$b_3 = 3\pi \cdot a$	$A_{sc\,3} = \frac{1}{2}\pi \cdot (3a)^2$	$A_3 = A_{sc\,3} - A_{sc\,1}$	$= 4\pi a^2$
$c_4(M_0, 4a)$	$b_4 = 4\pi \cdot a$	$A_{sc\,4} = \frac{1}{2}\pi \cdot (4a)^2$	$A_4 = A_{sc\,4} - A_{sc\,2}$	$= 6\pi a^2$
$c_5(M_u, 5a)$	$b_5 = 5\pi \cdot a$	$A_{sc\,5} = \frac{1}{2}\pi \cdot (5a)^2$	$A_5 = A_{sc\,5} - A_{sc\,3}$	$= 8\pi a^2$
$c_6(M_0, 6a)$	$b_6 = 6\pi \cdot a$	$A_{sc\,6} = \frac{1}{2}\pi \cdot (6a)^2$	$A_6 = A_{sc\,6} - A_{sc\,4}$	$= 10\pi a^2$
$c_7(M_u, 7a)$	$b_7 = 7\pi \cdot a$	$A_{sc\,7} = \frac{1}{2}\pi \cdot (7a)^2$	$A_7 = A_{sc\,7} - A_{sc\,5}$	$= 12\pi a^2$
$c_8(M_0, 8a)$ (above)	$b_8 = 8\pi \cdot a$	$A_{sc\,8} = \frac{1}{2}\pi \cdot (8a)^2$	$A_8 = A_{sc\,8} - A_{sc\,6}$	$= 14\pi \cdot a^2$
$c_8(M_0, 8a)$ (below)	$b_8 = 8\pi \cdot a$	$A_{sc\,8} = \frac{1}{2}\pi \cdot (8a)^2$	$A_9 = A_{sc\,8} - A_{sc\,7}$	$= \frac{15}{2}\pi a^2$

The length of the spiral is the sum of the b's

$$(1 + 2 + 3 + 4 + 5 + 6 + 7 + 8 + 8)\pi a = 44\,\pi a$$

We can test to see if we calculated the areas of the semi-annuli correctly by adding them to get the area of the largest circle.

$$A_1 + A_2 + A_3 + A_4 + A_5 + A_6 + A_7 + A_8 + A_9$$

$$= \frac{1}{2}\pi a^2 + 2\pi a^2 + 4\pi a^2 + 6\pi a^2 + 8\pi a^2 + 10\pi a^2 + 12\pi a^2 + 14\pi a^2 + \frac{15}{2}\pi a^2$$

$$= (\frac{1}{2} + 2 + 4 + 6 + 8 + 10 + 12 + 14 + \frac{15}{2})\,\pi a^2$$

$$= 64\,\pi a^2 = \pi(8a)^2 = Area_{circle\,8}$$

What is nice here is that with π's help we can calculate the length and area of the spiral.

The Unique Seven-Circles Arrangement

Try taking seven coins of the same size and placing them so that six of them are tangent to the seventh one, as shown below. (This can only be done with seven congruent circles.)

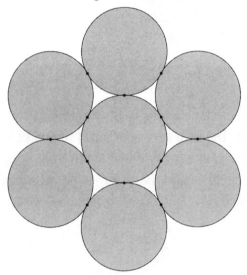

Fig. 6-16

You will discover as we go further into this section that the reason that this can only be done with seven congruent circles is that if you join the radii at the points of tangency, you will form a regular hexagon. This is analogous to drawing a circle with a pair of compasses and then finding out that if you mark off consecutively the radius length along the circumference of the circle, it will bring you back exactly to your starting point after six segments.

Consider the configuration in figure 6-17. We might want to determine the area of the nonshaded regions between the congruent circles. There are several ways of finding the area of the nonshaded regions. We will offer one here.

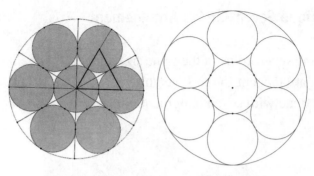

Fig. 6-17a   **Fig. 6-17b**

Consider the equilateral triangle surrounding one of these non-shaded regions. The area of a nonshaded region can be obtained by finding the area of the triangle and subtracting three sectors—each one-sixth of a circle (since it has a 60° angle). If we let the radius of a small circle equal r, we get one of the nonshaded regions as follows:

$$Area_{eq.\,triangle} = \frac{(2r)^2 \sqrt{3}}{4} = r^2 \sqrt{3}$$

$$Area_{shaded\,sectors} = 3\left(\frac{1}{6}\pi r^2\right) = \frac{\pi r^2}{2}$$

$$Area_{one\,nonshaded\,region} = r^2 \sqrt{3} - \frac{\pi r^2}{2} = \frac{r^2}{2}\left(2\sqrt{3} - \pi\right)$$

$$Area_{nonshaded\,regions} = 6\left[\frac{r^2}{2}\left(2\sqrt{3} - \pi\right)\right] = 3r^2\left(2\sqrt{3} - \pi\right)$$

To find the area of the six-pointed figure in the center (see fig. 6-18a), we merely add the area of one of the small circles to the sum of these nonshaded regions:

$$Area_{six\text{-}pointed\,figure} = \pi r^2 + 3\pi^2\left(2\sqrt{3} - \pi\right) = 2r^2\left(3\sqrt{3} - \pi\right)$$

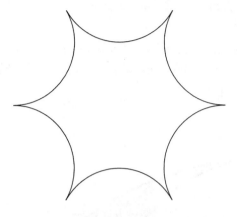

Fig. 6-18a

Fig. 6-18b

Once a circle is placed around the six outside circles, there are additional nonshaded regions. To get the total area of these non-shaded regions inside the larger circle, we simply subtract the total area of the seven small circles from the larger circle. $(3r)^2 \pi - 7\pi r^2 = 2\pi r^2$. Thus, with the help of π, we were able to show that the remaining area, when the seven circles are taken out of the larger circle, is the equivalent of two small circles.

A "Mushroom" Shape

The following figure consists of a quadrant (or quarter circle) and two overlaid semicircles, whose diameters are as big as the radius of the quarter circle.

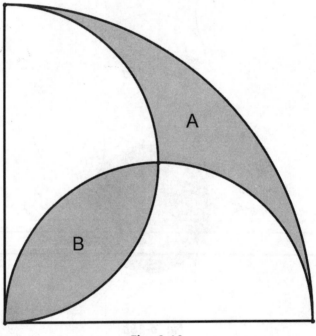

Fig. 6-19

What would you guess is the relationship between the shaded regions marked *A* and *B*?

The answer may become clearer if we complete the various circles.

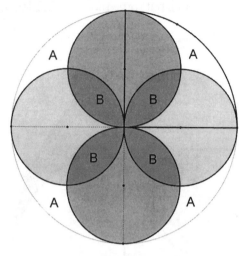

Fig. 6-20

Since the big circle has double the radius of the little circle, its area is four times as big.[9]

$$Area_{big\ circle} = 4 \cdot Area_{small\ circle} = 4(\pi r^2) = \pi(2r)^2$$

So the sum of the areas of the four inner circles is the same as the area of the outer circle.

We notice that there are four overlapping regions (marked *B*), and there are four regions in the larger circle that are not included in the smaller circles. Since the *B* regions are used twice and the *A* regions are not used at all in the sum of the area of the four smaller circles (and there is complete symmetry), we can conclude that each of the *B* regions must be equal in area to each of the *A* regions—recalling that the sum of the areas of the four smaller circles equals the area of the larger circle. This type of reasoning is very important in mathematics.

9. There is an important concept in geometry, namely, that two similar figures have areas in a ratio that is the square of their ratio of similitude (the ratio of their corresponding sides). This idea is used here since all circles are similar to each other.

We can also look at this problem in another way, one that may require a bit less abstraction, but some more work. Elegance has its price!

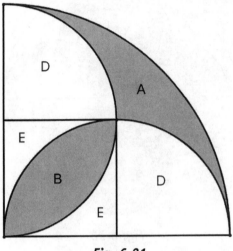

Fig. 6-21

Consider the figure above, where perpendicular radii of length *r* are drawn in the two semicircles. We can represent the various areas as follows:

First, to find the area of region *A*, we will subtract from the large quarter circle (which includes regions *A, D, D, E, E,* and *B*) the two smaller quarter circles and the small square (this includes regions *D, D, E, E,* and *B*).

$$Area_A = \frac{1}{4}\left(4\pi r^2\right) - \left[2\left(\frac{1}{4}\pi r^2\right) + r^2\right]$$

$$= \pi r^2 - \frac{\pi}{2}r^2 - r^2$$

$$= r^2\left(\frac{\pi}{2} - 1\right)$$

To find the area of region B, we add the two small (overlapping) quarter circles (this includes regions B, E, B, and E) and subtract the square (including regions E, E, and B) from this sum.

$$Area_B = 2\left(\frac{1}{4}\pi r^2\right) - \left(r^2\right)$$

$$= \frac{\pi}{2}r^2 - r^2$$

$$= r^2\left(\frac{\pi}{2} - 1\right)$$

So we can clearly see that the two regions A and B have the same area.

Over the next few pages we will be working on some unusual shapes. They will be formed by circle arcs inside a square. The following figures will foreshadow the ensuing discussion. Since it is said that a picture is equivalent to a thousand words, we will let these figures speak for themselves.

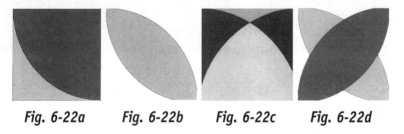

Fig. 6-22a   Fig. 6-22b   Fig. 6-22c   Fig. 6-22d

In figure 6-22a, the darker shaded region is a quarter circle of radius a. Therefore, to find this area we take one-quarter of the area of the circle. Thus, the area is $\frac{1}{4}\pi a^2$. To get the area of the lighter shaded region, we merely subtract the area of the quarter circle from the area of the square to get $a^2 - \frac{1}{4}\pi a^2 = a^2\left(1 - \frac{\pi}{4}\right)$. This technique will be used throughout this exploration of the areas of the strange regions we will be considering.

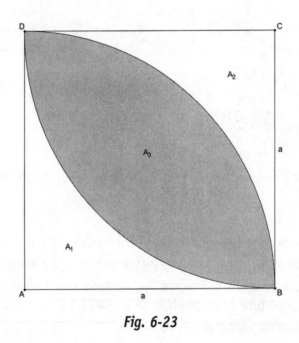

Fig. 6-23

In figure 6-23 (where *ABCD* is a square, and two quarter circle arcs *BD* are drawn), we are asked to find the area of the shaded region (the football shape)—comprised of two overlapping quarter circles. The straightforward way (which most people would probably use) is to find the area of sector *ADB* and subtract the area of right $\triangle ADB$, resulting in the area of the segment (half the football shape), which is then doubled to get the area of the shaded region.

A more elegant method (we believe) is to add the areas of sectors *ADB* and *CBD* to get the area equal to $A_1 + 2A_3 + A_2$. If we now subtract the area of the square from this sum, we get the area of the shaded region.

We will now carry out this plan:

Area sector $ADB = \frac{1}{4}\pi a^2$

Area sector $CBD = \frac{1}{4}\pi a^2$

Sum of areas of sectors $ADB + CBD = \frac{1}{2}\pi a^2$

[Notice that the shaded region is used twice in the addition.]

Subtract the area of square $ABCD$ to get $\frac{1}{2}\pi a^2 - a^2 = a^2\left(\frac{\pi}{2}-1\right)$

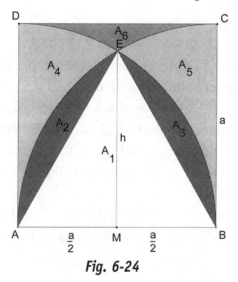

Fig. 6-24

In figure 6-24, we have two quarter circles centered at vertices A and B of square $ABCD$, with side length a. We seek to find the area of region A_6. The segment EM is perpendicular to $\overline{AB}$ at its midpoint M. $\triangle AEB$ is equilateral.

By the Pythagorean theorem, $EM = \frac{a\sqrt{3}}{2}$, so the area of $\triangle AEB$

$$= \frac{1}{2}\left(\frac{a\sqrt{3}}{2}\right)(a) = \frac{a^2\sqrt{3}}{4}.^{10}$$

---

10. This is the well-known and frequently used formula for finding the area of an equilateral triangle with side length given. In this case the side length is a.

Now to the solution of the problem: twice the area of sector *AEB* minus the area of $\triangle AEB$ gives the area of $A_1 + A_2 + A_3$.

Let's do that now.

Since $m \angle AEB = 60°$, the area of sector $AEB = \frac{\pi a^2}{6}$.

Double this is $\frac{\pi a^2}{3}$. Subtracting the area of $\triangle AEB$ gives us

$$\frac{\pi a^2}{3} - \frac{a^2 \sqrt{3}}{4}$$

We now have the area of $A_1 + A_2 + A_3 = \frac{\pi a^2}{3} - \frac{a^2 \sqrt{3}}{4}$.

We use a similar technique; however, this time we will find twice the area of quarter-wide sector *ADB* and subtract the region we just found: $A_1 + A_2 + A_3$. In other words, we again subtract the doubly used overlap region to get the shaded region. What is then left to complete the square is A_6, which is the region whose area we sought in the first place. The problem will then be solved.

Now for the computation:

The area of quarter circle sector $ADB = \frac{\pi a^2}{4}$, and double that is $\frac{\pi a^2}{2}$.

We must now subtract the overlapped region, used twice:

$$\frac{\pi a^2}{2} - \frac{\pi a^2}{3} + \frac{a^2 \sqrt{3}}{4}$$

And then subtract this from the area of the square to get

$$a^2 - \left(\frac{\pi a^2}{2} - \frac{\pi a^2}{3} + \frac{a^2 \sqrt{3}}{4} \right)$$

$$= a^2 - \frac{\pi a^2}{6} - \frac{a^2 \sqrt{3}}{4}$$

$$= a^2 \left(1 - \frac{\pi}{6} - \frac{\sqrt{3}}{4} \right), \text{ which is the area of } A_6.$$

We will be needing the area of this region (A_6) for the next problem, which may be a bit challenging. However, with the work we have already done and with the technique we have used a few times in these earlier problems—that of subtracting the overlapped region (used twice)—we should have no difficulty solving the problem.

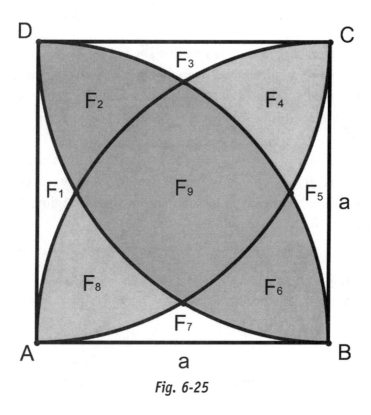

Fig. 6-25

In figure 6-25 we have our quarter circular sectors with centers at the vertices of square $ABCD$ and radius a, intersecting to form region F_9, whose area we seek. In the previous problem, we had just found the area of region F_3. We can get the area of the total shaded region, $F_2 + F_4 + F_6 + F_8 + F_9$, by subtracting the areas of the four unshaded regions (each equal to F_3) from the area of the square $ABCD$.

This is done as follows:

$$a^2 - 4\left[a^2\left(1 - \frac{\pi}{6} - \frac{\sqrt{3}}{4}\right)\right]$$

$$= a^2\left(\frac{2\pi}{3} - 3 + \sqrt{3}\right)$$

We can also get the area of this shaded region by finding the sum of the areas of the two overlapping "football"-shaped regions and subtracting the area of the overlap regions F_9 (which was used twice). We found the area of this football shaped region on page 183 to be $a^2(\frac{\pi}{2} - 1)$.

Twice that is $a^2(\pi - 2)$, which is the area of the shaded region plus the region F_9 (which was used twice). So, all we need to find the sought-after region is to subtract the shaded region, F_9, from twice the "football" region.

$$a^2(\pi - 2) - a^2\left(\frac{2\pi}{3} - 3 + \sqrt{3}\right)$$

$$= a^2\left(\frac{\pi}{3} + 1 - \sqrt{3}\right)$$

This was no mean feat. Yet you can see the role that π plays in these rather unusual excursions into finding the areas of strange regions.

A "Dolphin" Shape

The side of the square lattice (fig. 6-26) has a unit measure of a. We would like to find the perimeter and the area of this strange-looking shape, which we will call a dolphin shape.

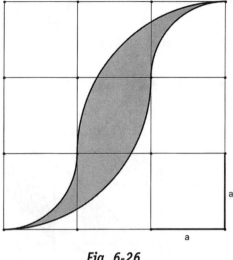

Fig. 6-26

Let's begin by inspecting the actual construction of the dolphin shape. For that we provide you with the complete circles, a part of which made up the dolphin shape. (See fig. 6-27.)

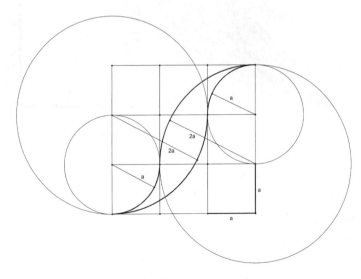

Fig. 6-27

The **perimeter** consists of two quarter circles with the radius $r_1 = a$ and two quarter circles with the radius $r_2 = 2a$. We get the perimeter rather easily in the following way.

$$Perimeter = 2\left(\tfrac{1}{4}\right)2\pi a + 2\left(\tfrac{1}{4}\right)2\pi(2a) = 3\pi a$$

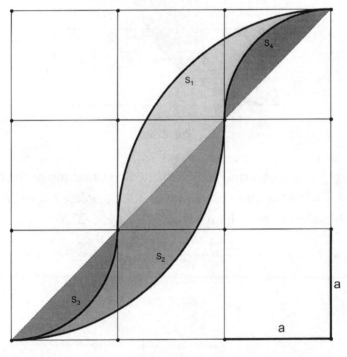

Fig. 6-28

The **area** consists of the areas of two segments with the radius $r_2 = 2a$, from which we subtract the areas of the two segments with the radius $r_1 = a$.

We can obtain the area of a segment (see fig. 6-29) by subtracting the area of the isosceles right triangle from the area of its quarter circle, that is, $Area_{segment} = Area_{quarter\ circle} - Area_{right\ triangle}$

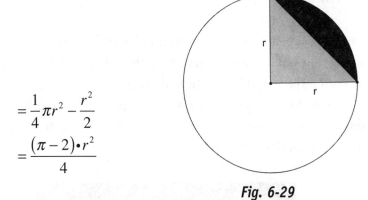

$$= \frac{1}{4}\pi r^2 - \frac{r^2}{2}$$

$$= \frac{(\pi - 2) \cdot r^2}{4}$$

Fig. 6-29

For the two segments S_1 and S_2 with the radius $r_2 = 2a$ we get the area

$$Area_{S_1} + Area_{S_2} = 2\left[\frac{(\pi - 2) \cdot (2a)^2}{4}\right] = 2\,(\pi - 2)a^2$$

then both segments S_3 and S_4 with the radius $r_1 = a$ deliver the area to be subtracted

$$Area_{S_3} + Area_{S_4} = 2\left[\frac{(\pi - 2) \cdot a^2}{4}\right] = \frac{1}{2}\,(\pi - 2)a^2$$

The area of the dolphin-shaped figure is then

$$\left(Area_{S_1} + Area_{S_2}\right) - \left(Area_{S_3} + Area_{S_4}\right) = 2(\pi - 2)a^2 - \frac{1}{2}(\pi - 2)a^2 = \frac{3}{2}(\pi - 2)a^2$$

You might like to discover other ways in which the perimeter and the area of this strange shape can be found.

The Yin and Yang

The yin-yang is an ancient symbol of Chinese philosophy. It reflects the polar basic concepts of the Chinese philosophy, from whose interplay and interaction all events of the universe arise.

Where does the yin-yang symbol come from?[11] The ☯ is a well-known Chinese yin-yang symbol. Its development is from the natural phenomena of our universe.

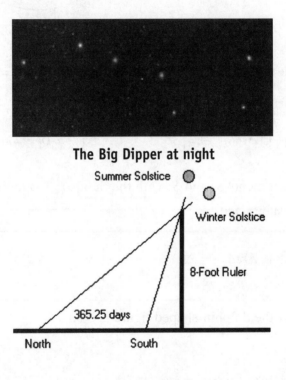

Fig. 6-30

By observing the sky, recording the Big Dipper's positions, and watching the shadow of the sun from an eight-foot (Chinese meas-

11. Parts of this section are taken from the Web site http://www.chinesefortunecalendar.com/yinyang.htm, which is copyright © 2003 by Allen Tsai. They are used with permission.

urement) pole, ancient Chinese people determined the four directions. The direction of sunrise is the east, the direction of sunset is the west, the direction of the shortest shadow is the south, and the direction of the longest shadow is the north. At night, the direction of the Polaris star is the north.

They noticed the seasonal changes. When the Big Dipper points to the east, it's spring; when the Big Dipper points to the south, it's summer; when the Big Dipper points to the west, it's fall; and when the Big Dipper points to the north, it's winter.

When observing the cycle of the sun, the ancient Chinese simply used a pole about eight feet long, posted at right angles to the ground, and recorded positions of the shadow. Then they found the length of a year is around 365.25 days. They even divided the year's cycle into twenty-four segments, including the vernal equinox, the autumnal equinox, the summer solstice, and the winter solstice, using the sunrise and Big Dipper positions.

They used six concentric circles, marked the twenty-four segment points, divided the circles into twenty-four sectors, and recorded the length of shadow every day. The shortest shadow is found on the day of the summer solstice. The longest shadow is found on the day of the winter solstice. After drawing lines and dimming the yin part going from the summer solstice to the winter solstice, the sun chart looks as it is shown below. The ecliptic angle 23° 26' 19" of the earth can be seen in the following chart.

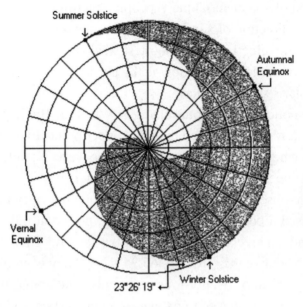

Fig. 6-31

The ecliptic is the sun's apparent path around the earth. It is tilted relative to the earth's equator. The value of obliquity of the ecliptic was around 23° 26' 19" in year 2000.

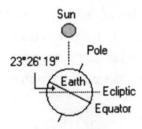

Fig. 6-32

By rotating the sun chart and positioning the winter solstice at the bottom, it will look like this ☯. The light-color area, which indicates more sunlight, is called yang (sun). The dark-color area has less sunlight (more moonlight) and is called yin (moon). Yang is like man. Yin is like woman. Yang wouldn't grow without yin. Yin couldn't give birth without yang. Yin is born (begins) at the summer solstice and yang is born (begins) at the winter solstice. Therefore, one little circle, yin, is marked on the summer solstice position. Another little circle, yang, is marked on the winter solstice position.

Fig. 6-33

In general, the yin-yang symbol is a Chinese representation of the entire celestial phenomenon. It contains the cycle of the sun through the four seasons.

The teardrop-looking symbol is bounded by three semicircles, where the two smaller ones each has a diameter half the length of the diameter of the larger semicircle.

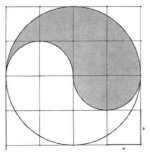

Fig. 6-34

To find the area of the yin-yang is simple, if we realize that what is taken away from the large semicircle, by the smaller overlapping semicircle, is added back to it again. So its area is simply that of the larger semicircle. (See fig. 6-35.)

To find the perimeter, we refer you back to chapter 1, where we established that the sum of the semicircular arcs inside another semicircle equals the larger semicircular arc (see pages 34–35). Therefore, since the sum of the lengths of the two smaller semicircular arcs equals the length of the larger semicircular arc, the yin-yang has a perimeter equal to the circumference of the larger circle. It may not appear that way, but it is so.

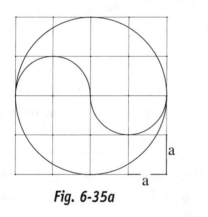

Fig. 6-35a

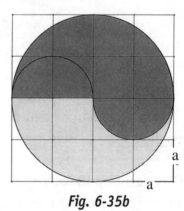

Fig. 6-35b

An alternative to the yin-yang is shown in the figure below (fig. 6-36). What is the area and perimeter of one of the "teardrop" shapes? And what is the area of the center region (that outside the teardrop shapes)?

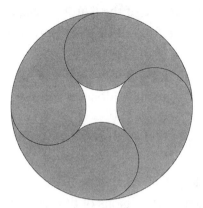

Fig. 6-36

As before, we shall assume symmetry, and as is indicated in figure 6-37, the radius of each small circle is r.

To find the perimeter of one of the teardrop shapes is easy, if we recognize that the small-circle portion of the teardrop shape is an arc of $180° + 45° = 225°$ of one of the small circles and an arc of $180° - 45° = 135°$ of another congruent small circle. Together that comprises the full $360°$ of a small circle. In addition, we need the arc length of one-quarter of the larger circle. To determine this, we first need to get the radius of the larger circle. With the Pythagorean theorem applied to right triangle BMG, we get $BM^2 + MG^2 = BG^2$, so that, because $BM = MG$, $BM = r\sqrt{2}$. The large circle has radius length $r + r\sqrt{2}$.

The computation is then straightforward:

The perimeter of the teardrop shape equals the circumference of the small circle plus one-quarter of the circumference of the larger circle.

$$\textbf{Perimeter} = 2\pi r + \frac{1}{4}\left[2\pi\left(r + r\sqrt{2}\right)\right] = \frac{\pi r}{2}\left(5 + \sqrt{2}\right)$$

To find the area of one of the teardrop shapes, we first will find the area of the center four-arc shape. Here we will focus on the square *BGEH*. The area of the center four-arc shape can be found by subtracting the areas of the four quarter circles from the area of the square *BGEH*. This can be done as follows:

$$\textbf{Area} = \left(2r\right)^2 - 4\left(\frac{1}{4}\pi r^2\right) = 4r^2 - \pi r^2 = r^2\left(4 - \pi\right)$$

Since the teardrop shape is comprised of one small circle and one of the four congruent outer regions of the large circle (which is not in the smaller circle), we simply subtract the areas of four small circles and the center four-arc shape from the area of the larger circle and take one-quarter of this result to get the area of one of these four "outer portions."

$\frac{1}{4}$[Area of larger circle – Area of center four-arc shape – Area of four smaller circles]

$$= \frac{1}{4}\left[\pi\left(r + r\sqrt{2}\right)^2 - r^2\left(4 - \pi\right) - 4\left(\pi r^2\right)\right]$$

$$= r^2\left(\frac{\pi\sqrt{2}}{2} - 1\right)$$

To finish finding the area of one of the teardrop shapes, we simply add this area to $\frac{225°}{360°} = \frac{5}{8}$ of the area of one of the smaller circles, which is $\frac{5}{8}\left(\pi r^2\right)$. Thus the area of one of the teardrop shapes is $r^2\left(\frac{\pi\sqrt{2}}{2} - 1\right) + \frac{5}{8}\left(\pi r^2\right) = r^2\left(\frac{4\pi\sqrt{2} + 5\pi - 8}{8}\right)$.

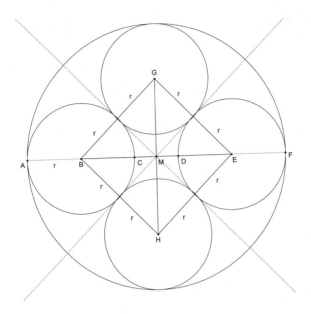

Fig. 6-37

Using π to Apportion a Pizza Pie Equally for Three People

With the task before us to divide a circular pizza equally among three people, we are faced with the problem of how to actually make the cuts. The pizza can be divided up evenly in many ways. We will show you four different ways here and challenge you to find other configurations that will yield three equal pieces of pizza.

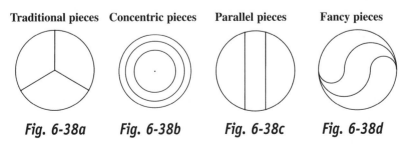

Traditional pieces	Concentric pieces	Parallel pieces	Fancy pieces
Fig. 6-38a	*Fig. 6-38b*	*Fig. 6-38c*	*Fig. 6-38d*

Traditional Pieces

If somebody would actually go to a pizza restaurant with a pro-
tractor, he would draw laughs from others and yet be able to draw
the appropriate radii that would partition the circle into three equal
parts, each of which is $\frac{360°}{3} = 120°$ —if the cheese cooperates. One
can also bring a piece of string to the restaurant and place it around
the pizza. Dividing the string into three parts, he would get each
piece as one-third of the circumference so

$$b = \frac{C}{3} = \frac{2\pi \cdot r}{3}$$

Everybody gets an equally big slice (sector).

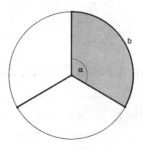

Fig. 6-39

Concentric Circles

Of course this concentric division is hardly suitable in the
restaurant, yet this version, from a mathematical point of view,
is quite interesting.

The radius r (of the initial circle c) is given. We must determine the two other radii, r_1 and r_2, so that the inner circle will be equal in area to the two annuli.

Given: radius of the larger circle, r

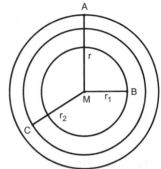

Sought-after: r_1 and r_2, to make the three areas equal

Solution: $r_1 = \frac{\sqrt{3}}{3} r$; $r_2 = \frac{\sqrt{6}}{3} r$

Fig. 6-40

Our objective is to find the values of r_1 and r_2 in terms of r so that
$$Area_1 = Area_{annulus\ BC} = Area_{annulus\ AC}$$

We begin with $Area_1 = \pi\, r_1^2$

To get the inside annulus, we subtract the areas of the two circles making up the annulus:
$$Area_{annulus\ BC} = Area_2 - Area_1 = \pi\, r_2^2 - \pi\, r_1^2 = \pi(\, r_2^2 - r_1^2\,)$$

We repeat this procedure for the outside annulus:
$$Area_{annulus\ AC} = Area - Area_2 = \pi r^2 - \pi\, r_2^2 = \pi(r^2 - r_2^2\,)$$

For the three regions (or pieces) to be equal in area, the following must be true:
$$\pi\, r_1^2 = \pi(\, r_2^2 - r_1^2\,) = \pi(r^2 - r_2^2\,)$$

Then by dividing through by π, the following is arrived at:

$$r_1^2 = r_2^2 - r_1^2 = r^2 - r_2^2$$

$r_1^2 = r_2^2 - r_1^2$, which gives us that $2\,r_1^2 = r_2^2$, and so $r_2 = r_1\,\sqrt{2}$

$r_2^2 - r_1^2 = r^2 - r_2^2$ enables us to get $2\,r_2^2 - r_1^2 = r^2$

Then substituting for r_2, we get

$4\,r_1^2 - r_1^2 = r^2$, and so $3\,r_1^2 = r^2$, or $r_1 = \frac{r\sqrt{3}}{3}$

and consequently, $r_2 = r_1\,\sqrt{2} = \sqrt{2}\left(r\frac{\sqrt{3}}{3}\right) = r\frac{\sqrt{6}}{3}$

Fancy Pieces (Using the Teardrop Shape)

We will use semicircles with the diameter trisected (i.e., divided into three equal segments).

Now that you have had some experience with this shape, you can see below how this will be divided into semicircular arcs.

Given: r

Sought-after: r_1 and r_2, so that $\overline{AB}$ is trisected

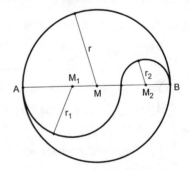

Solution: $r_1 = \frac{2}{3}r$ and $r_2 = \frac{1}{3}$

Fig. 6-41

We will first find the area ($Area_{teardrop}$) of the teardrop shown above:

$$r_1 = \frac{2}{3}r \text{ and } r_2 = \frac{1}{3}r$$

$$Area_{SC(M_1)} = \frac{1}{2}\pi r_1^2 = \frac{1}{2}\pi \left(\frac{2}{3}r\right)^2 = \frac{2}{9}\pi r^2$$

$$Area_{SC(M_2)} = \frac{1}{2}\pi r_2^2 = \frac{1}{2}\pi \left(\frac{1}{3}r\right)^2 = \frac{1}{18}\pi r^2$$

$$Area_{SC(M)} = \frac{1}{2}\pi r^2$$

$$Area_{teardrop} = Area_{SC(M)} - Area_{SC(M_1)} + Area_{SC(M_2)} = \frac{1}{2}\pi r^2 - \frac{2}{9}\pi r^2 + \frac{1}{18}\pi r^2$$

$$Area_{teardrop} = \frac{1}{3}\pi r^2$$

If this teardrop has an area of one-third of the circle, then clearly the other one (above the horizontal diameter) also must be one-third the area of the circle. So, if these two teardrops encompass two-thirds of the area of the circle, then the remaining portion in the middle must also be one-third of the area of the circle.

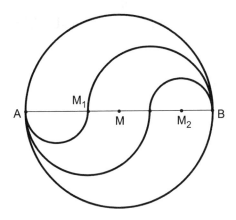

Fig. 6-42

Parallel Pieces

If one cannot find the midpoint of the pizza (see the first method), cutting parallel pieces offers us another approach, but it will get more difficult than expected! (*Caution:* the average reader may want to just look at the result, since the method is a bit complicated.)

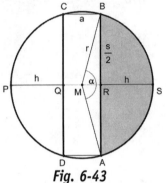

Given: Circle with radius r

Sought-after: α or h

Solution: $\alpha \approx 149.3°$ or $h \approx \frac{3}{4} r$

Fig. 6-43

It should be clear that if we are looking for the point at which to make our two cuts, namely, $\overline{AB}$ and $\overline{CD}$, we would have to know the length of h or the measure of α.

In the triangle $\triangle MRB$ we get

$$\sin\frac{\alpha}{2} = \frac{\frac{s}{2}}{r} = \frac{s}{2r}; \text{ therefore, } \frac{s}{2} = r\sin\frac{\alpha}{2}, \text{ and } MR^2 = MB^2 - RB^2$$

we then get

$$(r-h)^2 = r^2 - (\tfrac{s}{2})^2 = r^2 - r^2\sin^2\tfrac{\alpha}{2} = r^2(1 - \sin^2\tfrac{\alpha}{2}) = r^2\cos^2\tfrac{\alpha}{2} \qquad (*)$$

$$\frac{Area_{sector(\alpha)}}{Area_{circle}} = \frac{\alpha}{360°}, \text{ and this can be written as}$$

$$Area_{sector\,(\alpha)} = \frac{\alpha}{360°} \cdot Area_{circle} = \frac{\alpha}{360°} \cdot \pi r^2$$

Using the relationship marked (*), above, we get

$$Area_{\Delta ABM} = \frac{1}{2} AB \bullet MR = \frac{1}{2} s \, (r - h) = \frac{1}{2} \bullet 2r \sin \frac{\alpha}{2} \bullet r \cos \frac{\alpha}{2}$$

$$= r^2 \sin \frac{\alpha}{2} \bullet \cos \frac{\alpha}{2}$$

With the double angle formula ($\sin 2x = 2\sin x \cos x$), this allows us to simplify further[12]

$$Area_{\Delta ABM} = \frac{1}{2} r^2 \bullet \sin \alpha$$

$$Area_{segment} = Area_{sector\,(\alpha)} - Area_{\Delta ABM} = \frac{\alpha}{360°} \bullet \pi \bullet r^2 - \frac{1}{2} r^2 \bullet \sin \alpha$$

$$= (\frac{\alpha}{360°} \bullet \pi - \frac{1}{2} \bullet \sin \alpha)r^2$$

The latter shall be a third of the circular area A, so that

$$(\frac{\alpha}{360°} \bullet \pi - \frac{1}{2} \bullet \sin \alpha)r^2 = \frac{1}{3} \pi r^2, \text{ which simplifies to}$$

$$\frac{\alpha}{360°} \bullet \pi - \frac{1}{2} \bullet \sin \alpha = \frac{1}{3} \pi, \text{ yielding } \sin \alpha = \frac{2\pi\alpha}{360°} - \frac{2}{3} \pi.$$

This transcendental equation cannot be solved in the traditional way. As an approximation, we can use a calculator and get $\alpha = 149.2741654...° \approx 149.3°$.

With $r - h = r \cos \frac{\alpha}{2}$, we then finally get $h = r \bullet (1 - \cos \frac{\alpha}{2})$

$$\approx 0.7350679152r, \text{ which is almost } h \approx \frac{3}{4} r.$$

We have thus divided the circle into three equal parts in four different ways. Can you find another method for trisecting the circle?

---

12. We can also get this directly if one knows the formula for the area of a triangle: $A = \frac{1}{2} a \, b \sin \gamma$.

The Constant Ring

There are times when π plays a somewhat minor role. It is actually upstaged by some very elegant techniques. This is the case with the following. A chord of the big circle touches the smaller circle in exactly one point (i.e., it is tangent to the smaller circle). We are given the length *s* of this chord, and we are asked to find the area of the shaded region (the ring).

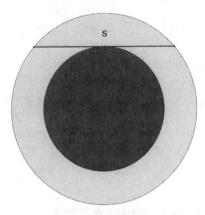

Fig. 6-44

You would think that there is not enough information given here, or is it possible to formulate this described situation in a variety of ways? The following figures show the segment *AB* maintaining its constant length, and yet the area between the circles (the ring) takes on a very different appearance. It would appear, by observation, that each configuration would yield a different area between the circles. Surprisingly, that is not the case. They all have the same area, as we shall see.

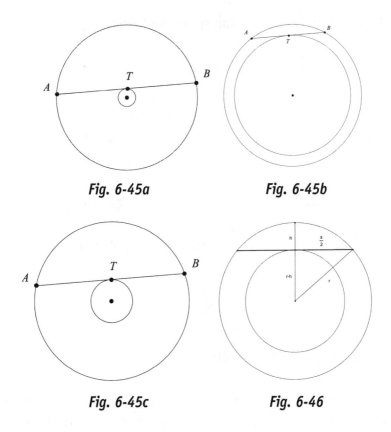

Fig. 6-45a   Fig. 6-45b

Fig. 6-45c   Fig. 6-46

By the Pythagorean theorem:

$$r^2 = (r - h)^2 + \left(\frac{s}{2}\right)^2 = r^2 - 2rh + h^2 + \left(\frac{s}{2}\right)^2$$

Therefore, $\left(\frac{s}{2}\right)^2 = 2rh - h^2$

$$Area_{large\ circle} = \pi r^2$$

$$Area_{small\ circle} = \pi \cdot (r - h)^2 = \pi \cdot r^2 - 2\pi rh + \pi h^2$$

$$Area_{ring} = Area_{outside} - Area_{inside} = \pi r^2 - (\pi r^2 - 2\pi rh + \pi h^2)$$
$$= 2\pi rh - \pi h^2 = \pi(2rh - h^2)$$

From this equation, we can substitute to get

$$Area_{ring} = \pi\left(\frac{s}{2}\right)^2$$

So the area of the ring only actually depends on the length s of the chord—and, of course, π plays its usual helpful role!

Although the diameter of neither circle was given, the problem shall nevertheless have a unique solution. We can therefore let the smaller circle become very small, so small that it has essentially zero length.

The ring then consists only of the larger (or outer) circle, whose diameter is then the chord ($2r = s$).

The ring area can be calculated now simply with the equation for the area of a circle:

$$Area = \pi r^2 = \pi\left(\frac{s}{2}\right)^2$$

The Constant Ring Extended

We shall consider another situation that can be viewed as an outgrowth of the ring area. We take once again a chord of specific (constant) length, and this time we will not have a concentric circle inside the larger circle, rather one that is tangent to the larger circle and, of course, as before, tangent to the given segment.

We draw a circle c with the radius r, such that a given (line) segment of length t is a chord of this circle (fig. 6-48).

Furthermore, we draw, respectively, the inner touching circles (with the radii r_1 and r_2).

What is the area of the lightly shaded region, which is formed by the six semicircular arcs shown in the figure?

At first, we notice that there are, of course, many different possibilities to draw such a circle c, as seen in the following figures.[13]

---

13. Note the connection between this situation and the arbelos (p. 211).

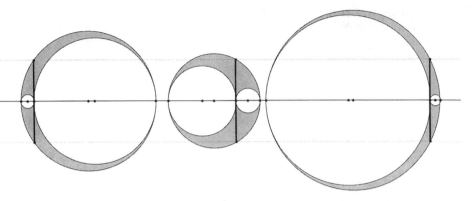

Fig. 6-47

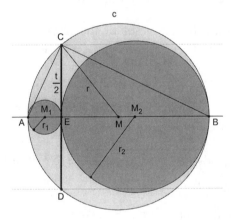

Fig. 6-48

With the Pythagorean theorem, there are various ways to estab-lish the relationship between the tangent segment and the radii of the two circles.

Pythagorean theorem applied to right $\triangle EMC$	Mean proportional of the altitude to the hypotenuse of right $\triangle ABC$	The product of the segments of the two intersecting chords
$MC^2 = ME^2 + CE^2$	$CE^2 = AE \bullet BE$	$CE \bullet DE = AE \bullet BE$
$r^2 = (r - 2r_1)^2 + \left(\frac{t}{2}\right)^2$	$\left(\frac{t}{2}\right)^2 = 2r_1 \bullet 2r_2$	$\left(\frac{t}{2}\right)^2 = 2r_1 \bullet 2r_2$
$r^2 = r^2 - 4rr_1 + 4r_1^2 + \left(\frac{t}{2}\right)^2$	$\frac{t^2}{4} = 4 \bullet r_1 \bullet r_2$	$\frac{t^2}{4} = 4 \bullet r_1 \bullet r_2$
$t^2 = 16r_1(r - r_1)$, yet	$t^2 = 16 \bullet r_1 \bullet r_2$	$t^2 = 16 \bullet r_1 \bullet r_2$
$r - r_1 = r_2$ (explained below)		
$t^2 = 16 \bullet r_1 \bullet r_2$		

This leads in all three cases to

$$t^2 = 8 \bullet 2r_1 \, r_2, \text{ which can be written as } \frac{t^2}{8} = 2r_1 \, r_2.$$

We know that the diameter of the largest circle can be represented as $2r = 2r_1 + 2r_2$, so we get $r = r_1 + r_2$, or written another way, $r_2 = r - r_1$.

We can find the area of the lightly shaded region by subtracting the areas of the two smaller circles from the area of the largest circle. We do this as follows:

$$Area_{lightly \, shaded \, region} = Area_{largest \, circle \, M} - Area_{circle \, M_1} - Area_{circle \, M_2}$$

$$= \pi \, r^2 - \pi \, r_1^2 - \pi \, r_2^2 = \pi(r^2 - r_1^2 - r_2^2) = \pi[r^2 - r_1^2 - (r - r_1)^2]$$

$$= \pi(-2 \, r_1^2 + 2r \bullet r_1) = 2\pi r_1(r - r_1) = 2\pi r_1 \bullet r_2$$

But from above we already established that $\frac{t^2}{8} = 2r_1 \, r_2$, so we then substitute to get

$$Area_{lightly \, shaded \, region} = Area_{largest \, circle \, M} - Area_{circle \, M_1} - Area_{circle \, M_2} = \pi \frac{t^2}{8},$$

which indicates that the area is independent of the radius r of the circle c.

The lightly shaded circular arc figure always has the same area if we put any circle through the ends of the (line) segment with the length t. Compare this to the arbelos on page 211.

The Lost Circle Area

Suppose you have four equal pieces of string. With *the first piece of string*, one circle is formed. *The second piece of string* is cut into two equal parts, and two congruent circles are formed. *The third piece of string* is cut into three equal pieces, and three congruent circles are formed. In a similar way, four congruent circles are formed from *the fourth piece of string*.

They are shown in figure 6-49. Note that the sum of the circumferences of each group of congruent circles is the same (since we used the same length string for each set of circles).

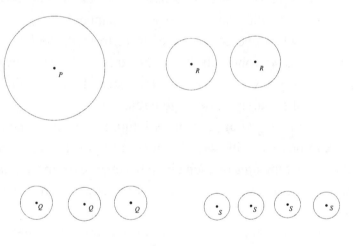

Fig. 6-49

Circle	Diameter	Each circle's circumference	Sum of the circles' circumferences	Each circle's area	Sum of the circles' areas	Percent of area of circle P represented by the sum of the areas of smaller circles
P	12	12π	12π	36π	36π	100
R	6	6π	12π	9π	18π	50
Q	4	4π	12π	4π	12π	$33\frac{1}{3}$
S	3	3π	12π	2.25π	9π	25

An inspection of the above chart shows that the sum of the circumferences for each group of circles is the same, yet the sum of the areas is quite different. The more circles we formed with the same total length of string, the smaller the total area of the circles. Just what you would likely *not* expect to happen!

That is, when two equal circles were formed, the total area of the two circles was one-half that of the large circle. Similarly, when four equal circles were formed, the total area of the four circles was one-fourth of the area of the large circle.

This seems to go against one's intuition. Yet if we consider a more extreme case, with, say, one hundred smaller equal circles, we would see that the area of each circle becomes extremely small, and the *sum* of the areas of these one hundred circles is one-hundredth of the area of the larger circle.

Try to explain this rather disconcerting concept. It ought to give you an interesting perspective on comparison of areas.

What would happen if the circles made from this piece of string were not of equal size? Try to use the above argument to see that you would end up with an analogous result.

Unusual Circle Relationships

The concept of π is neatly embedded in computation with circles. Sometimes it merely plays an accompanying role, as is the case with some fascinating circle relationships that have been known to us for over two thousand years.

Archimedes came up with some rather astonishing geometric phenomena regarding circles. They show an intuitive facility with the concept of π, even if he didn't have it calculated as accurately as we have today.

We offer two such examples just to fascinate you with some unusual circle relationships.

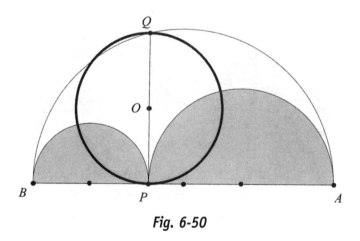

Fig. 6-50

The *arbelos*, or shoemaker's knife, is obtained by drawing three semicircles, two along the diameter of the third. The two smaller semicircles can be of any size, as long as their diameters encompass the entire diameter of the third semicircle. Thus, $AP + PB = AB$. What Archimedes said is that the area between the semicircles (unshaded) is equal to the area of the circle with $\overline{PQ}$ as diameter.

Notice that $\overline{PQ}$ is the perpendicular segment from P, the point of intersection of the two smaller semicircles, to Q, the point where the perpendicular meets the largest semicircle.

This can be easily proved. We only need to remember a theorem from elementary geometry, namely, that the altitude to the hypotenuse of a right triangle is the mean proportional between the two segments along the hypotenuse. That is, PQ is the mean proportional between AP and PB, or $\frac{AP}{PQ}=\frac{PQ}{PB}$. With the radii of the two smaller semicircles having lengths a and b, respectively, this gives us

$$h^2 = 2a \cdot 2b = 4ab, \text{ so } h = 2\sqrt{ab}$$

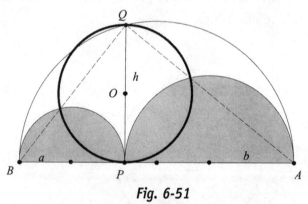

Fig. 6-51

First, we shall represent the area of the region between the semicircles. This is done by finding the area of the largest semicircle and subtracting from that area the areas of the two smaller semicircles.

$$\begin{aligned}
\text{Area of unshaded region} &= \frac{\pi}{2}(a+b)^2 - \frac{\pi}{2}a^2 - \frac{\pi}{2}b^2 \\
&= \frac{\pi}{2}\left(a^2 + 2ab + b^2 - a^2 - b^2\right) \\
&= \frac{\pi}{2}(2ab) \\
&= \pi ab
\end{aligned}$$

The area of the circle with diameter PQ and radius $\frac{h}{2}$ is

$$\pi\left(\tfrac{h}{2}\right)^2 = \pi\left(\sqrt{ab}\right)^2 = \pi ab$$

Both areas are the same.

Another ingenious relationship that Archimedes discovered and published is called *Salinon*. It states that the area bounded by the four semicircles (in fig. 6-52a), where $AB = EF$, is equal to the area of the circle with the diameter PS, where PS is the perpendicular line segment through R and having endpoints on the two semicircles. The next few figures will demonstrate this with a variety of different arrangements.

In each case, compare the area of the lightly shaded region to that of the circle.

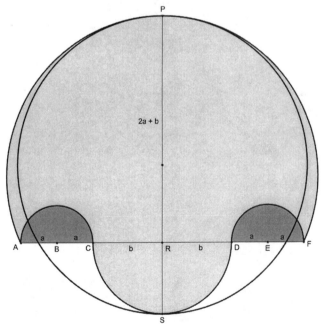

Fig. 6-52a

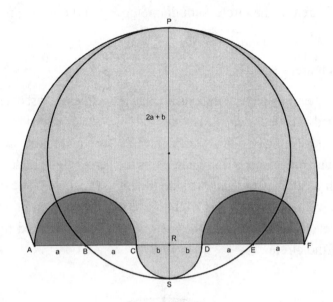

Fig. 6-52b

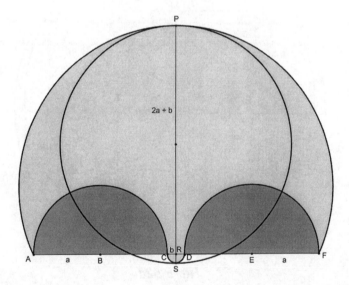

Fig. 6-52c

Notice that in the third of these variations, we are approaching a special case of the arbelos (fig. 6-52 a,b, and c), when the little semicircle at the bottom almost disappears.

To prove this, we subtract the area of the two equal semicircles with radii $EF = AB = a$ from the large semicircle with diameter AF. We then add the area of the semicircle with radius $CR = b$.

$$\left[\frac{1}{2}\pi(2a+b)^2 - \pi a^2\right] + \frac{1}{2}\pi b^2$$

$$= \left[\frac{1}{2}\pi(4a^2 + 4ab + b^2) - \pi a^2\right] + \frac{1}{2}\pi b^2$$

$$= \pi(a^2 + 2ab + b^2)$$

$$= \pi(a + b)^2$$

The area of the circle with radius $a + b$ is $\pi(a + b)^2$, and so they are equal, despite the relative size of the semicircles. These are two truly wonderful relationships, since they are independent of the relative sizes. Just imagine to have discovered these relationships without the tools and experience we have today.

π and the Imaginary Unit *i*

After all is said and done, π also has a role in mathematics to help explain certain concepts. You may recall that an imaginary number is one that includes $i = \sqrt{-1}$. However, is i^i also an imaginary number? To answer this question, we need π. The mathematically curious reader should be able to follow the proof below. For those not so inclined, suffice it to say, we are able to show (with the help of π) that i^i is a real number, and not an imaginary number, as might have been suspected.

The proof follows:

With $x = \frac{\pi}{2}$, in $e^{ix} = \cos x + i \cdot \sin x$ we obtain

$$e^{i \cdot \frac{\pi}{2}} = \cos \frac{\pi}{2} + i \cdot \sin \frac{\pi}{2} = 0 + i = i,$$ which yields

$i = e^{i \cdot \frac{\pi}{2}}$, which yields

$$i^{\,i} = (e^{i \cdot \frac{\pi}{2}})^i = e^{-\frac{\pi}{2}} = \frac{1}{\sqrt{e^{\pi}}} \approx 0.207879576351\ldots$$

Therefore, $i^{\,i}$ is a real number!

In 1746 the famous mathematician Leonhard Euler showed that $i^{\,i}$ adopts infinitely many values, all of which are real.

For example, $i^{\,i} = e^{-\left(\frac{\pi}{2} + 2k\pi\right)}$ with $k \in \mathbf{Z}$ ($\mathbf{Z}$ = set of all integers);

when $k = 0$: $i^{\,i} = (e^{i \cdot \frac{\pi}{2}})^i = e^{-\frac{\pi}{2}} = \frac{1}{\sqrt{e^{\pi}}} \approx 0.207879576351\ldots$

We have now gone through a wide variety of applications of π. Some were of the real-life variety, and others made use of the circle's applications. You now realize that π can be seen as a number with unusual properties or as the ratio that defines it. In the latter case, we are talking about circles. In the former case, we see the constant interrelationships between seemingly unrelated concepts in mathematics surfacing to the fore.

Chapter 7

Paradox in π

A paradox is a seemingly contradictory statement that may nonetheless be true. In geometry paradoxes appear in many forms. The following is one such example. Consider four congruent circular objects (fig. 7-1a) tied together by an elastic string. Then shift the circles to the position shown in figure 7-1b. In which case is the elastic string longer?

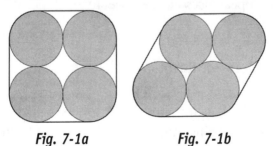

Fig. 7-1a   Fig. 7-1b

If we look at figures 7-2a and 7-2b, we notice that in each figure the elastic string is comprised of four line segments, each equal to the diameter of the circles. Therefore, the only comparison needed is that of the lengths of the circular arcs.

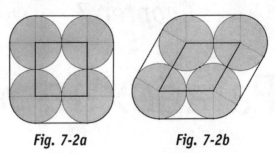

Fig. 7-2a   **Fig. 7-2b**

In figure 7-2a, the four arcs are each one-quarter circle; hence the remaining elastic (four quarter arcs) is that of the circumference of the circle (we would need our trusty π to determine its length). The arcs of elastic string in figure 7-2b are each supplementary to one of the angles of the rhombus[1] in the middle of the figure. Yet the sum of the angles of the rhombus (as for any quadrilateral) is 360°. Therefore the sum of the elastic string arcs must also be a full 360°, and the remaining elastic string length is that of the circumference of the circle. Lo and behold, the two pieces of elastic string have the same length. Appearances can be deceptive!

Rolling Cylinders—π Revolutions!

Heavy loads are often transported on rollers, which are not connected to the object being transported. What might be the advantage to using rollers that are not connected to the transported object?

---

1. A rhombus is a quadrilateral with four equal sides. A square is a special type of rhombus.

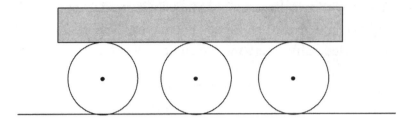

Fig. 7-3

How far does the transported object moved to the right, if the rollers, which have a diameter of 1 foot, have made one revolution on their own axes? One would expect that the object has moved the distance of one revolution, or the length of the circumference of the circle. In this case, π feet.

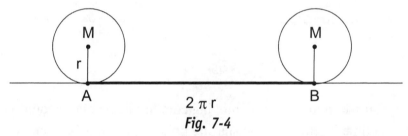

Fig. 7-4

The easiest, and perhaps most elegant way, to explain the movement of the object on the rollers is to think of the rollers moving π feet in one revolution, and the object also moves π feet with respect to the rollers. Therefore, the object moves 2π feet with respect to the ground. We simply add the two distances.

This becomes more complex (yet analogous) when we consider two congruent circular disks. Consider one disk rolling around the other disk. How many revolutions will the moving disk make as it travels once around the stationary disk? You will probably guess that since the circumferences are equal, the moving disk will have made one revolution. Wrong! It makes two revolutions.

Try it with two large coins. Mark their starting positions and then notice how many revolutions the moving coin has made when it has traveled halfway around the stationary coin.

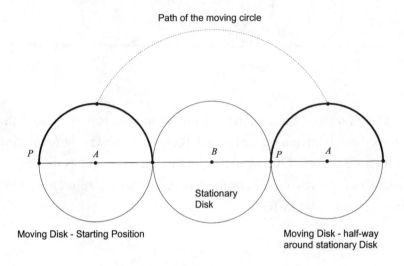

Fig. 7-5

You will notice that the moving coin has made two revolutions by the time it returns to the starting point of the stationary circle.

A Constant among Concentric Circles

The now-famous ratio of the circumference to the diameter of a circle, π, shows itself nicely as a constant relating two or more concentric circles.

Consider the following problem:

Two concentric circles are ten units apart, as shown below (fig.7-6). What is the difference between the circumferences of the circles?

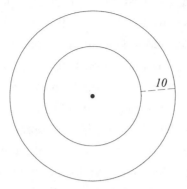

Fig. 7-6

The traditional straightforward method for solving this problem is to find the diameters of the two circles. Then, finding the circumference of each circle, we would merely have to subtract to find their difference. Since the lengths of the diameters are not given, the problem is a bit more complicated than that. Let d represent the diameter of the smaller circle. Then $d + 20$ is the diameter of the larger circle.[2] The circumferences of the two circles will then be πd and $\pi(d + 20)$, respectively.

The difference of the circumferences is $\pi(d + 20) - \pi d = 20\pi$.

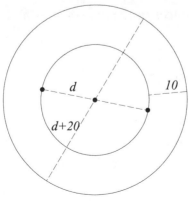

Fig. 7-7

---

2. That is, the diameter, d, of the smaller circle plus twice 10, the distance between the circles.

A more elegant, and vastly more dramatic, procedure would be to use an extreme case. To do this, we will let the smaller of the two circles become smaller and smaller until it reaches an "extreme smallness" and becomes a "point."[3] In this case, the circle shrunk to a point would become the center of the larger circle. The distance between the two circles now becomes simply the radius of the larger circle. The difference between the lengths of the circumferences of the two circles at the start is now merely the circumference of the larger circle,[4] or 20π.

Although both procedures yield the same answer, notice how much more work is used for the traditional solution by actually taking the difference of the lengths of the circumferences of the two circles, and how using the idea of considering an extreme situation (without compromising any generality), we reduced the problem to something relatively easy.

We could also look at the problem without being "distracted" by the concentric placement of the circles. We are looking for the differences of the circumferences $C_2 - C_1$, where $C_1 = \pi d_1$ and $C_2 = \pi d_2$. So that $C_2 - C_1 = \pi d_2 - \pi d_1 = \pi(d_2 - d_1)$. Verbally expressed, we have shown that the difference of the circumferences is equal to π times the difference of the diameters. Another formulation would be that the ratio of the differences of the circumferences to the differences of the diameters is π.

A Rope around the Equator

We are about to embark on an amazing paradox. We will establish a "fact" you will probably find intuitively impossible to accept. Before we begin, let's assume that the earth is a perfect sphere—

---

3. We can do this because we weren't given the size of either circle, so as long as we preserve the length 10, we can consider the two circles to take on any convenient sizes.

4. Since the smaller circle has a circumference of 0.

just to make our work a bit easier. We begin by placing an imaginary rope (tautly) around the equator of the earth. Assume also that the earth is a smooth surface along the equator. We will now lengthen the rope by exactly one meter. The rope is now loose. Let's situate the rope so that it is everywhere equidistant from the surface of the earth. Our question is: Can a mouse easily fit in the space between the rope and the surface of earth? What do you think? The answer will certainly surprise you.

Fig. 7-8

We are not the inventors of this problem. Apparently the first publication of this "classic" problem is contained in the article "The Paradox Party. A Discussion of Some Queer Fallacies and Brain-Twisters" by Henry Ernest Dudeney:[5]

> Mr. Smoothly, the curate, at the end of the table, said at this point that he had a little question to ask.
>
> "Suppose the earth were a perfect sphere with a smooth surface, and a girdle of steel were placed round the equator so that it touched at every point."

5. *The Strand Magazine. An Illustrated Monthly*, ed. George Newnes, 38, no. 228 (December 1909): 670–76; *Amusements in Mathematics* (London: Thomas Nelson and Sons, 1917; reprint, New York: Dover, 1970), p. 139.

"I'll put a girdle round about the earth in forty minutes," muttered George, quoting the words of Puck in *A Midsummer Night's Dream*.

"Now, if six yards were added to the length of the girdle, what would then be the distance between the girdle and the earth, supposing that distance to be equal all round?"

"In such a great length," said Mr. Allgood, "I do not suppose the distance would be worth mentioning."

"What do you say, George?" asked Mr. Smoothly.

"Well, without calculating I should imagine it would be a very minute fraction of an inch."

Reginald and Mr. Filkins were of the same opinion.

"I think it will surprise you all," said the curate, "to learn that those extra six yards would make the distance from the earth all round the girdle very nearly a yard!"

"Very nearly a yard!" everybody exclaimed, with astonishment; but Mr. Smoothly was quite correct. The increase is independent of the original length of the girdle, which may be round the earth or round an orange; in any case the additional six yards will give a distance of nearly a yard all round. This is apt to surprise the nonmathematical mind.

As we begin to tackle this problem, we will assume that the earth is a perfect sphere,[6] and, for the sake of simplicity, we will assume that the equator is exactly 40,000 kilometers long. We are even going to be more extreme than the story above, in that we will only add one meter to the length of the rope.

Before beginning, what would you guess the answer to be? Remember we are extending the 40,000-kilometer rope, which is taut around the equator, so that it is now 40,000.001 kilometers long and is now placed equidistant above the equator. If you are doubtful about a mouse fitting under this rope, would you think we could slide a pencil under this rope?

---

6. In reality the earth is a geoid, not a perfect sphere.

Let's consider the figure below (fig. 7-9).

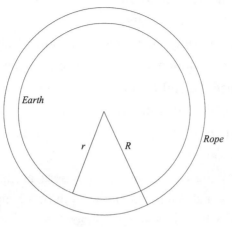

Fig. 7-9

The familiar circumference formulas give us

$$C = 2\pi r, \text{ or } r = \frac{C}{2\pi}$$

and

$$C + 1 = 2\pi r, \text{ or } R = \frac{C+1}{2\pi}$$

We need to find the difference of the radii, which is

$$R - r = \frac{C+1}{2\pi} - \frac{C}{2\pi} = \frac{1}{2\pi} \approx .159 \text{ m} \approx 16 \text{ cm}$$

Can you imagine, there is actually a space of about 16 centimeters between the rope and the earth's surface all the way around? So there is more than enough space (about 16 cm) for a mouse to crawl beneath it.

You must really appreciate this astonishing result. Imagine, by lengthening the 40,000-kilometer rope by 1 meter, it lifted off the equator about 16 centimeters!

Consider the original problem diagramed above. You should realize that the solution was independent of the circumference of the earth, since the end result did not include the circumference in its final calculation. It required only calculating $\frac{1}{2\pi}$. Here again you see how π stays in the picture even when the dimensions of the circle have disappeared.

Instead of the earth, we could choose an apple, a Ping-Pong ball, or even a disk such as a dollar coin or a penny.

A thread, which is 1 meter longer than the apple's circumference (coin's circumference), is concentrically wound around an apple (or a coin). What distance, a, is the thread from the apple's surface (or, for that matter, from the coin's edge)?

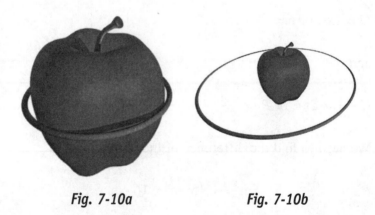

Fig. 7-10a   **Fig. 7-10b**

Let r equal the radius of the apple, and $r + a$ is then the radius of the thread (at 1 meter longer).

(1) $l_{thread} = C_{apple} + 1 = 2\pi r + 1$

(2) $l_{thread} = 2\pi(r + a)$

With $2\pi r + 1 = 2\pi(r + a) = 2\pi r + 2\pi a$, we immediately again get $1 = 2\pi a$; this yields $a = \frac{1}{2\pi}$. As expected, we get the same results as before $a = 0.1591549430... \approx 0.159$ m ≈ 16 cm, once more emphasizing that in this situation the result is independent of the apple's radius.

This independence of the radius or circumference (of the earth, an apple, or a Ping-Pong ball) is particularly confirmed here.

The distance a is only dependent on the chosen extension (1 m)—and naturally on our trusty π.

You might feel compelled to carry out such an experiment once (for example, with a coin, a Ping-Pong ball, and a basketball); this then convinces the skeptics, too.

We can take advantage of this independence further by solving this problem with the useful problem-solving technique of considering extreme values. Suppose we reduce the original circle as far as is possible. Let's go even further so that it reduces to a point. Then the length of the radius of the circle of the rope is now what we are to find and is quite easily obtained.

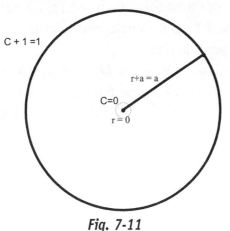

Fig. 7-11

The result nevertheless remains the same. In our problem, the extension of 1 meter is itself the circle's circumference, and its radius is the sought-after distance, a.

Using this technique, we suppose the inner circle (above) is very small, so small that it has a zero-length radius (that means it is actually just a point). We are required to find the difference between the radii, $R - r = R - 0 = R$.

So all we need to find is the length of the radius of the larger circle and our problem will be solved. With the circumference of the smaller circle now 0, we apply the formula for the circumference of the larger circle:

$$C + 1 = 0 + 1 = 2\pi R, \text{ so } R = \frac{1}{2\pi}$$

This problem has two lovely little treasures. First, it reveals an astonishing result, clearly not to be anticipated at the start, and, second, it provides you with a nice problem-solving strategy that can serve as a useful model for future use.

The result sometimes causes people to repeat their calculations—to see if they really have misjudged the original problem, or if they really can't trust their intuition. This is for many people a paradox because the result is independent of the earth's radius. But the result is dependent on π.

This paradoxical result can also be formulated in the following way:

For the difference of the circumferences of two concentric circles with the radii R and r and the distance a between the circumferences, we get

$C_1 - C_2 = 2\pi R - 2\pi r$
But $R = r + a$
Therefore, $C_1 - C_2 = 2\pi R - 2\pi r = 2\pi(r + a) - 2\pi r$
$= 2\pi r + 2\pi a - 2\pi r = 2\pi a$

Notice the similarity of the "formula" for the difference of the circumferences and the formula for the circumference of a circle. Both are dependent on π.

To provide a better (or deeper) understanding of this unusual resulting dependency on π, consider the following diagram, where each of the two circumferences is "rolled out" to form a straight line.

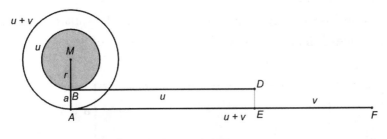

Fig. 7-12

It may be helpful (because of $2\pi \approx 6$) to consider the circumference of a circle as about six times as long as the accompanying radius to understand the issue at hand. We may then regard the radius as one-sixth of the circumference. Therefore, the radius $\overline{MA}$ is one-sixth of the length of circumference $\overline{AF}$, and $\overline{MB}$ is one-sixth of the length of $\overline{BD}$. Consequently, the difference, a (or AB), between the two radii is also one-sixth of the difference EF of the circumferences.

The length of $\overline{AB}$ is only dependent on the difference between the two circumferences, not on the lengths of the respective circumferences. However big (or small) AF and BD may be, when the difference EF is exactly 1 meter long, then the difference AB between the two radii is about $\frac{1}{6}$ meter long, or about 17 centimeters. This also applies when the lengths BD and AF are represented by the equator and the rope is 1 meter longer.

Suppose we chose a square instead of the (equator) circle. We can examine an analogous situation that may shed further light on this unusual situation.[7]

---

7. This proposal was made by Heinrich Winter, *Entdeckendes Lernen* (Wiesbaden/ Braun-schweig: Vieweg, 1991), p. 163.

A rope in the form of a square is placed around a square. The perimeter of the square rope is 1 meter longer than that of the original square and placed so that the sides of both squares are parallel and equidistant all around.

What is this distance, *a*, that the rope is from the square's sides? This is a question analogous to that of the circle before.

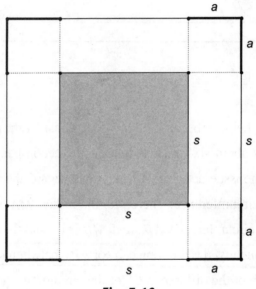

Fig. 7-13

	side	perimeter
Initial square	*s*	4*s*
Rope square	*s* + 2*a*	4(*s* + 2*a*)

The diagram shows clearly that the extra 1 meter rope length is accounted for by eight equally long pieces, which overlap at the corners, and that they are exactly as long as the distance between the parallel sides of the two squares. Therefore, the distance between the parallel sides of the squares must be $\frac{1}{8}$ of 1 m = 0.125 m = 12.5 cm.

The distance, a, between the squares is also independent of the size of the initial square. It is merely one-eighth of the difference of the perimeters of the rope and the square. Remember that the distance between the two concentric circles before was also a constant $\left(\frac{1}{2\pi}\right)$ times the difference of the perimeters (i.e., circumferences). How are these two constants analogous? Let's consider similar situations for other regular polygons.

Instead of a square, one could also use an equilateral triangle or, in general, any regular polygon, and seek to find the distance between the 1-meter-lengthened rope and the sides of the polygon.

Around a regular polygon a rope is placed that is 1 meter longer than the perimeter of the polygon. The rope is shaped into a similar polygon and placed so that the respective polygon sides are parallel to the rope polygon.

What is the distance, a, that the rope is from the sides of the polygon? More precisely, what is the distance between the parallel sides of the two polygons? This will vary with the number of sides of the regular polygon. Look at the following results. (A detailed calculation can be found in appendix D.)

For a three-sided regular polygon (an equilateral triangle)

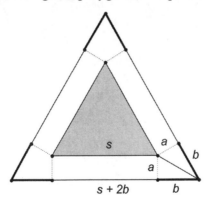

Fig. 7-14
$a \approx 0.096 \text{ m} = 9.6 \text{ cm}$

For a four-sided regular polygon (a square)

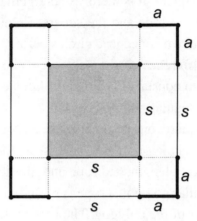

Fig. 7-15
$a = 0.125$ m = 12.5 cm

For a regular pentagon

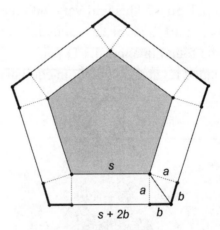

Fig. 7-16
$a \approx 0.138$ m = 13.8 cm

For a regular hexagon

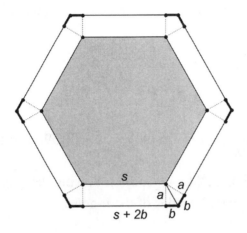

Fig. 7-17
$a \approx 0.144$ m $= 14.4$ cm

Notice how the value of a, the distance of the 1-meter-lengthened rope from the regular polygon, increases as the number of sides of the regular polygon increases. What would you expect will be the maximum value of a? The maximum number of sides for a regular polygon could be considered a polygon that is like a circle. So we would expect that a will get gradually larger until it reaches the value we had for the circle, about 15.9 centimeters.

The greater the number of sides of the regular polygon, the larger the distance, a, that separates the two similar polygons. Does a increase infinitely? Of course, this distance, a, can never become larger than that for a circle.

As the number of sides, n, gets infinitely large, we obtain for the perimeter (in this case the circumference of a circle) the limiting value $C = 2r$. Now π reappears again.

In the case of the regular hexagon, the distance a (14.4 cm) already lies relatively near the limit value (which is approximately 15.9 cm) that we obtained from the circle earlier (rope around the equator):

$$\lim_{n \to \infty} \frac{1}{2n \tan \dfrac{\pi}{n}} = \frac{1}{2\pi} = 0.15915494\ldots$$

and $a = \dfrac{1}{2\pi} = 0.1591549430 \ldots \approx 0.159$ m ≈ 16 cm

For $n = 3, 4, \ldots, 24$ we get the following distances between the corresponding parallel sides of the n-gons:

n	exact value for a	approximate value for a
3	$\dfrac{\sqrt{3}}{18}$	0.096225044
4	$\dfrac{1}{8}$	0.125
5	$\sqrt{\dfrac{\sqrt{5}}{250} + \dfrac{1}{100}}$	0.13763819
6	$\dfrac{\sqrt{3}}{12}$	0.14433756
7	$\dfrac{\cot \dfrac{\pi}{7}}{14}$	0.14832295
8	$\dfrac{\sqrt{2}}{16} + \dfrac{1}{16}$	0.15088834
9	$\dfrac{\cot \dfrac{\pi}{9}}{18}$	0.15263763
10	$\sqrt{\dfrac{\sqrt{5}}{200} + \dfrac{1}{80}}$	0.15388417

11	$\dfrac{\cot\dfrac{\pi}{11}}{22}$	0.15480396
12	$\dfrac{\sqrt{3}}{24} + \dfrac{1}{12}$	0.15550211
13	$\dfrac{\cot\dfrac{\pi}{13}}{26}$	0.15604459
14	$\dfrac{\cot\dfrac{\pi}{14}}{28}$	0.15647450
15	$\dfrac{\cot\dfrac{\pi}{15}}{30}$	0.15682100
16	$\sqrt{\dfrac{\sqrt{2}}{512}+\dfrac{1}{256}} + \dfrac{\sqrt{2}}{32} + \dfrac{1}{32}$	0.15710435
17	$\dfrac{\cot\dfrac{\pi}{17}}{34}$	0.15733904
18	$\dfrac{\cot\dfrac{\pi}{18}}{36}$	0.15753560
19	$\dfrac{\cot\dfrac{\pi}{19}}{38}$	0.15770188
20	$\sqrt{\dfrac{\sqrt{5}}{800}+\dfrac{1}{320}} + \dfrac{\sqrt{5}}{40} + \dfrac{1}{40}$	0.15784378
21	$\dfrac{\cot\dfrac{\pi}{21}}{42}$	0.15796586
22	$\dfrac{\cot\dfrac{\pi}{22}}{44}$	0.15807165

23	$\dfrac{\cot\dfrac{\pi}{23}}{46}$	0.15816392
24	$\dfrac{\sqrt{2}}{48} + \dfrac{\sqrt{3}}{48} + \dfrac{\sqrt{6}}{48} + \dfrac{1}{24}$	0.15824487

It becomes clear why the distance a, in the case relating to a circle, escapes our immediate intuitive understanding. The topic of infinity is perhaps at fault: the 1-meter extension to the rope (around the equator) must be cut into an "infinite number" of parts if we are to follow the polygon models just considered.

We can have some fun with this concept of independence of circle size—relying only on our friend, π. This time, instead of placing a rope around the equator, 16 centimeters above the surface of the sphere as we did before, we shall construct an imaginary railway track around the equator. However, the inner rail will touch the surface of the equator, while the outer rail is suspended in the air above the equator (perpendicular to the surface).

How many meters longer would the outer rail be than the inner rail of such a railway line round the equator, if the inner rail is exactly 40,000,000 meters long?

We will let *a* represent the distance between the two rails. Here *a* = 1.46 meters.

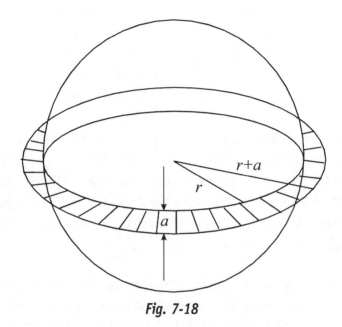

Fig. 7-18

From previous examples, you might already be able to predict what the outcome here will be. To what extent will the answer be dependent on the length of the equator?

We know that the circumference (*c*) of the equator is (for our purposes) 40,000,000 meters, and the distance between the rails (*a*) is 1.46 meters. We seek to find the difference of the two circumferences of the rails: *C* − *c*, where *C* is the circumference of the outer rail.

$C - c = 2\pi(r + a) - 2\pi r = 2\pi a$, which you will again notice is independent of the size of the two circles. To find the difference of the two rail lengths, we simply multiply 2π times 1.46, the value of *a*. This is 9.1734505484821962316 . . . ≈ 9.17 meters. The result would be the same (9.17 m) if we chose any other sphere instead of the earth. That may be hard to accept, but our trust in π always remains faithful to us, regardless!

An analogous situation arises if you were to walk along the equator (40,000,000 m) and ask: How much further will your head travel than your feet, if you are 1.8 meters (about 5'11") tall?

The famous novelist Jules Verne has one of his characters trying to calculate which body parts travel farther during a world tour— the head or the feet. This is what we are after here.

From the previous example, we can see that the answer is again independent of the length of the walk. Rather, it is dependent on π. We need merely multiply 2π times the person's height (in this case 1.8 meters) to get the answer: 11.30973355292325552 . . . ≈ 11.31 meters.

So for the trek along the equator, the head will have traveled almost 11.5 meters farther than the feet. That the result is independent of the (earth's) radius becomes more obvious when the following situation is considered.

A man, who is 1.80 meters tall, walks once around the earth's equator and also around a cylindrical space flight capsule (circumference 20 m) in space. In both cases, his head has traveled a greater distance than his feet. How much longer would the distance traveled by his head be when he walks around the equator, as compared with when he walks around the space flight capsule?

Let's consider the extreme case, that the man's feet are attached to a rotatable axle and he does a complete circle around this axle with his body stretched out. We must find the distance that the head has traveled, while for our purposes his feet have traveled an almost-zero distance.

Similarly to our earlier problem, we consider the almost-zero distance the feet traveled to be zero. So that all we need to find is the circumference of a circle of radius 1.8 meters. That is, $C = 2\pi r = 2\pi(1.8) \approx 11.31$ m.

This might also be seen in an "upside-down" version, namely, a trapeze artist holding on to a bar with his hands and spinning around—his hands now take the place of his feet, and his feet take the place of his head.

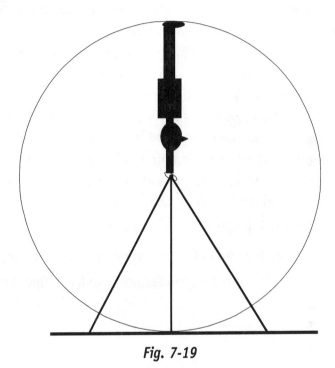

Fig. 7-19

In all these cases, we notice that the size of the circle is not the critical issue; rather, only the value of π gives us what we seek. This should let you appreciate even more the power of π.

Another Surprise

Now that your intuition has been somewhat tarnished by the surprising results of the rope around the earth, we present another possible situation. The rope that is 1 meter longer than the circumference of the earth is now no longer spaced out over the equator. Rather, it is pulled taut from one external point. Remember when the rope was equally spaced above the equator, there was merely a space of 16 centimeters. Now you will be surprised. The 1-meter-longer rope pulled taut from a point, where the rest of the rope "hugs" the earth's surface, reaches a point about 122 meters above the earth's surface.

Let's see why this is so. This time the answer is clearly dependent on the size of the earth and not exclusively on π—but remember π will also play a role here.

From the exterior point T, the rope (1 meter longer than the circumference of the equator) is pulled taut so that it hugs the earth's surface until it determines the points of tangency (S and Q). We seek to find how high up from the surface the point T is. That means we will try to find the length of x or $\overline{RT}$.

Remember the length of the rope from B through S to T is 0.5 meter longer than the circumference of the earth. So that the lengths of $\overparen{BS} + \overline{ST} = \overparen{BSR} + 0.5$ m. We are going to try to find the length of $\overline{TR}$ (or x).

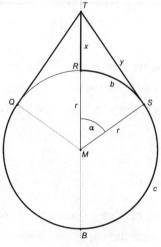

Fig. 7-20

So let's review where we are: the rope lies on the arc $\overparen{SBQ}$, which ends in the points S and Q and at points S and Q goes tangential to the point T. The lengths in the figure above are marked, and $\alpha = m\angle RMS = m\angle RMQ$.

The length of the rope $+ 1 = 2\pi r + 1$, and we get the following relationships:

$y = b + 0.5$. This is equivalent to $b = y - 0.5$ (y is 0.5 m longer than b, because of the extension by 1 m).

In ΔMST, the tangent function will be applied: $\tan \alpha = \frac{y}{r}$, and $y = r \cdot \tan \alpha$.

We can form the ratio of arc length to central angle measure and get the following:

$$\frac{b}{\alpha} = \frac{2\pi r}{360°} \text{ , and then we can get } b = \frac{2\pi \cdot r \cdot \alpha}{360°}$$

With $C = 2\pi r$, we can find the earth's radius (assuming that the equator is exactly 40,000,000 meters long).

$$r = \frac{C}{2\pi} = \frac{40,000,000}{2\pi} \approx 6,366,198 \text{ m}$$

Combining the equations we have above, we get the following:

$$b = \frac{2\pi \cdot r \cdot \alpha}{360°} = y - 0.5 = r \cdot \tan \alpha - 0.5$$

We are now faced with a dilemma, namely, that this equation (obtained above) cannot be uniquely solved in the traditional manner. We will set up a table of possible trial values to see what will fit (i.e., satisfy the equation).

$$\frac{2\pi \cdot r \cdot \alpha}{360°} = r \cdot \tan \alpha - 0.5$$

We will use the value of r we found above: $r = 6{,}366{,}198$ meters.

α	$b = \dfrac{2\pi \bullet r \bullet \alpha}{360°}$	$b = r \bullet \tan \alpha - 0.5$	Comparison of values (number of places in agreement—**bold**)
30°	3,333,333.478	3,675,525.629	1
10°	**1,1**11,111.159	**1,1**22,531.971	2
5°	**55**5,555.5796	**55**6,969.6547	2
1°	**111,1**11.1159	**111,1**21.8994	4
0.3°	**33,333.**33478	**33,333.**13940	5
0.4°	**44,444.4**4637	**44,444.4**6844	5
0.35°	**38,888.89**057	**38,888.89**7430	6
0.355°	**39,444.44**615	**39,444.44**5091	6
More exactly			
0.353°	**39,222.22**392	**39,222.22**019	7
0.354°	**39,333.335**04	**39,333.335**54	**8**
0.3545°	**39,388.89**059	**39,388.89**322	7
0.355°	**39,444.44**615	**39,444.44**5091	6

Our various trials would indicate that our closest match of the two values occurs at $\alpha \approx 0.354°$.

For this value of α, $y = r \bullet \tan \alpha \approx (6{,}366{,}198)(0.006178544171)$ $\approx 39{,}333.83554$ meters, or about 39,334 meters.

The rope is therefore almost 40 kilometers long before it reaches its peak. But how high off the earth's surface is the rope? That is, what is the length of x?

Applying the Pythagorean theorem to ΔMST, we get $MT^2 = r^2 + y^2$.

$MT^2 = (6{,}366{,}198)^2 + (39{,}334)^2 = 40{,}528{,}476{,}975{,}204 + 1{,}547{,}163{,}556 = 40{,}530{,}024{,}138{,}760$

So $MT \approx 6{,}366{,}319.512$ m

We are looking for x, which is $MT - r \approx 121.5120192$ m, or about 122 meters.

This result is perhaps astonishing because one intuitively assumes that by the circumference of the earth (40,000 km), an extra meter must almost disappear. But this is the mistake! The larger the sphere, the farther the rope can be pulled away form it.

Looking at the extreme case, where the radius of the equator decreases to zero, we have the minimum value for x, namely, $x = 0.5$ meter.

Thus, we have seen that π can also play a role in "fooling" us, or teasing our intuition. What this tells us is that the ratio of a circle's circumference to its diameter is a very special number in mathematics. So from now on, you should never take that number, called π, for granted. It should have earned a place in your mathematical mind as a very special number to behold.

Epilogue

By now you can confidently say that you know what π is. It emanated from the constant ratio of the circumference of a circle to its diameter. The four thousand years of struggles by the world's greatest mathematicians to establish its exact value have resulted in a mere approximation of its value, which at the time of publication is an accuracy to 1.24 trillion decimal places. Despite not establishing its exact decimal equivalent, we are able to use the concept in a plethora of applications, some of which we presented in this book. The concept of π also provided us with curiosities and other forms of mathematical entertainment that should serve as motivation for the reader to seek out further properties of this ubiquitous number π.

Now that we have considered many aspects of this most notable number, it is appropriate to actually see it a bit better than before. So here is π to one hundred thousand decimal places. Enjoy!

π=3.14159265358979323846264338327950288419716939937510582097494459230781640628 6
20899862803482534211706798214808651328230664709384460955058223172535940812848 11
17450284102701938521105559644622948954930381964428810975665933446128475648233 7
86783165271201909145648566923460348610454326648213393607260249141273724587006 6
06315588174881520920962829254091715364367892590360011330530548820466521384146 9
51941511609433057270365759591953092186117381932611793105118548074462379962749 56
73518857527248912279381830119491298336733624406566430860213949463952247371907 0
21798609437027705392171762931767523846748184676694051320005681271452635608277 8
57713427577896091736371787214684409012249534301465495853710507922796892589235 4
20199561121290219608640344181598136297747713099605187072113499999983729780499 51
05973173281609631859502445945534690830264252230825334468503526193118817101000 3
13783875288658753320838142061717766914730359825349042875546873115956286388235 3
78759375195778185778053217122680661300192787661119590921642019893809525720106 54
85863278865936153381827968230301952035301852968995773622599413891249721775283 4
79131515574857242454150695950829533116861727855889075098381754637464939319255 0
60400927701671139009848824012858361603563707660104710181942955596198946767837 4
49448255379774726847104047534646208046684259069491293313677028989152104752162 0
56966024058038150193511253382430035587640247496473263914199272604269922796782 3
54781636000934172164121992458631503028618297455570674983850549458858692699569 09
27210797509302955321165344987202755960236480665499119881834797753566369807426 54
25278625518184175746728909777727938000816470600161452491921732172147723501414 4
19735568548161361157352552133475741849468438523323907394143334547762416862518 98
35694855620992192221842725502542568876717904946016534668049886272327917860857 8
43838279679766814541009538837863609506800642251252051173929848960841284886269 4
56004241965285022210661186306744278622039194945047123713786609563643719172874 6
77646575739624138908658326459958133904780275900994657640789512694683983525957 0
98258226205224894077267194782684826014769909026401363944374553050680234962524 5
17493996514314298091906592509372216964615157098583874105978859597729754989301 6
17539284681382686838689427741559918559252459539594310499725246808459872736446 9
58486538367362226260991246080512438843904512441365497627807977156914359977001 2
96160894416948668555848406353422072225828488684158456028506016842739452267467 67
88952521385225499546667278239864565961163548862305774564980355936345681743241 12
51507606947945109659609040252288797108931456691368672287489405601015033086179 28
68092087476091782493858900971490967598526136554978189312978482168299894872265 8
80485757564014270477555132379641451523746234364542858444795265867821051141354 735
73952311342716610213596953623144295248493718711014576540359027993440374200731 05
78539062198387447808478489683321445713868751943506430218453191048481005370614 6
80674919278191197939952061419663428754440643745123718192179988391015919561814 6
75142691239748940907186494231961567945208095146550225231603881930142093762137 8
55956638937787083039069792077346722182562599661501421503068038447345492026054 14
66592520149744285073251866600213243408819071048633173464965145390579626856100 55
08106658796998163574736384052571459102897064140110971206280439039759515677157 7
00420337869936007230558763176359421873125147120532928191826186125867321579198 4
14848829164470609575270695722091756711672291098169091528017350671274858322287 1
18352093539657251210835791513698820914442100667510334671103141267111369908658 5
16398315019701651511685171437657618351556508849099898599823873455283316355076 4

7918535893226185489632132933089857064204675259070915481416549859461637180270981994309924488957571282890592323326097299712084433573265489382391193259746366730583604142813883032038249037589852437441702913276561809377344403070746921120191302033038019762110110044929321516084244485963766983895228684783123552658213144957685726243344189303968642624341077322697802807318915441101044682325271620105265227211166039666557309254711055785376346682065310989652691862056476931257058635662018558100729360659876486117910453348850346113657686753249441668039626579787718556084552965412665408530614344431858676975145661406800700237877659134401712749470420562230538994561314071127000407854733269939081454664645880797270826683063432858785698305235808933065757406795457163772542021149557615814002501262285941302164715509792592309907965473761255176567513575178296664547791745011299614890304639947132962107340437518957359614589019389713111790429782856475032031986915140287080859904801094121472213179476477726224142548545403321571853061422881375850430633217518297986622371721591607716692547487389866549494501146540628433663937900397692656721463853067360965712091807638327166416274888800786925602902284721040317211860820419000422966171196377921337575114959501566049631862947265473642523081770367515906735023507283540567040386743513622224771589150495309844489333096340878076932599397805419341447377441842631298608099888687413260472156951623965864573021631598193195167353812974167729478672422924654366800980676928238280689964004824354037014163149658979409243237896907069779422362508221688957383798623001593776471651228935786015881617557829735233446042815126272037343146531977774160319906655418763979293344195215413418994854447345673831624993419131814809277771038638773431772075456545322077709212019051660926804909263601975988281613323166636528619326686336062735676303544776280350450777235547105859548702790814356240145171806246436267945612753181340783303362542327839449753824372058353114771199260638133467768796959703098339130771098704085913374641442822772634659470474587847787201927715280731767907707157213444730605700733492436931138350493163128404251219256517980694113528013147013047816437885185290928545201165839341965621349143415956258658655705526904965209858033850722426482939728584783163057777560688876446248246857926039535277348030480290058760758251047470916439613626760449256274204208320856611906254543372131535958450687724602901618766795240616342522577195429162991930645537799140373404328752628889639958794757291746426357455254070914513571113694109119393251910760208252026187985318870584297259167781314969900901921169717372784768472686084900337702424291651300500516832336435038951702989392233451722013812806965011784408745196012122859937162313017114448464090389064495444006198690754851602632750529834918740786680881833851022833450850486082503930213321971551843063545500766828294930413776552793975175461395398468339363830474611996653858153842056853386218672523340283087112328278921250771262946322956398989893582116745627010218356462201349671518819097303811980049734072396103685406643193950979019069963955245300545058068550195673022921913933918568034490398205955100226353536192041994745538593810234395544959778377902374216172711117236434354394782218185286240851400666044332588856986705431547069657474585503323233421073015459405165537906866273337995851156257843229882737231989875714159578111963583300594087306812160287649628674460477464915995054973742526901049037781986835938146574126804925648798556145372347867330390468838343634655379498641927056387293174872332083760112302991136793862708943879936201629515413371424892e8

307220126901475466847653576164773794675200490757155527819653621323926406160136
358155907422020203187277605277219005561484255518792530343513984425322341576233
610642506390497500865627109535919465897514131034822769306247435363256916078154
781811528436679570611086153315044521274739245449454236828860613408414863776 7009
612071512491404302725386076482363414334623518975766452164137679690314950191085
759844239198629164219399490723623464684411739403265918404437805133389452574239
950829659122850855582157250310712570126683024029295252201187267675622041542051
618416348476516999811614101002996078386909291603028840026910414079288621 50784
245167090870006992821206604183718065355672525325675328612910424877618258297651
579598470356222629348600341587229805349896502262917487882027342092222453398562
647669149055628425039127577102840279980663658254889264880254566101729670266407
655904290994568150652653053718294127033693137851786090407086671149655834343476
933857817113864558736781230145876871266034891390956200993936103102916161528813
843790990423174733639480457593149314052976347574811935670911013775172100 8031559
024853090669203767192203322909433467685142214477379393751703443661991040337 5111
735471918550464490263655128162288244625759163330391072253837421821408835086573
917715096828874782656995995744906617583441375223970968340800535598491754173818
839994469748676265516582765848358845314277568790029095170283529716344562129640
435231176006651012412006597558512761785838292041974844236080071930457618932349
229279650198751872127267507981255470958904556357921221033346697499235630254947
802490114195212382815309114079073860251522742995818072471625916685451 3331239480
494707911915326734302824418601426363954800044800267049624820179289647669 75831
832713142517029692348896276684403232609275249603579964692565049368183609003238
092934595889706953653494060340216654437558900456328822505452556405644824651518
754711962184439658253375438856909411303150952617937800297412076651479394259 0298
969594699556576121865619673378623625612521632086286922210327488921865436480229
678070576561514463204692790682120738837781423356282360896320806822246801224826
117718589638140918390367367222088832151375560037279839400415297002878307667094
447456013455641725437090697939612257142989467154357846878861444581231459357198
492252847160504922124247014121478057345510500801908699603302763478708108175450
119307141223390866393833952942578690507643100638351983438934159613185434754649
556978103829309716465143840700707360411233759984345225161050702705623526601276
484830840761183013052793205427462865403603674532865105706587488225698157936789
766974220575059683440869735020141020672358502007245225632653413410559240190 27421
624843914035998953539459094407046912091409387001264560016237428802109276457931
065792295524988727584610126483699989225695968815920560010165525637567856672279
661988578279484885583439751874455512965634434803966420557982936804352202770 98
429423253302257634180703947699415979159453006975214829336655566156787364005366
656416547321704390352132954352916941459904160875320186837937023488868947915107
163785290234529244077365949563051007421087142613497459561513849871375704710178
795731042296906667021449863746459528082436944578977233004876476524133907592043
401963403911473202338071509522010682563427471646024335440051521266932493 41967
397704159568375355516673027390074972973635496453328886984406119649616277344951
827369558822075735517665158985519098666539354981068873206859907540792342 40230
092590070173196036225475647894064754834664776041146323390565134330684495397907
090302346046147096169688688850140834704054607429586991382966824681857103 1887906

5287036650832431974404771855678934823089431068287027228097362480939962706074726455399253994428081137369433887294063079261595995462624629707062594845569034711972996409089418059534393251236235508134949004364278527138315912568989295196427287573946914272534366941532361004537304881985517065941217352462589548730167600298865925786628561249665523533829428785425340483083307016537228563559152534784459818313411290019992059813522051173365856407826484942764411376393886924803118364453698589175442647399882284621844900877697763127957226726555625962825427653183001340709223343657791601280931794017185985999338492354956400570995585611349802524990669842330173503580440811685526531170995708994273287092584878944364600504108922669178352587078595129834417295351953788553457374260859029081765155780390594640873506123226112009373108048548526357228257682034160504846627750450031262008007998049254853469414697751649327095049346393824322271885159740547021482897111777923761225788734771881968254629812686858170507402725502633290449762778944236216741191862694396506715157795867564823993917604260176338704549901761436412046921823707648878341968968611815581587360629386038101712158552726683008238340465647588040513808016336388742163714064354955618689641122821407533026551004241048967835285882902436709048871181909094945331442182876618103100735477054981596807720094746961343609286148494178501718077930681085469000944589952794243981392135055864221964834915126390128038320010977386806628779239718014613432445726400973742570073592100315415089367930081699805365202760072774967458400283624053460372634165542590276018348403068113818551059797056640075094260878857357960373245141467867036880988060971642584975951380693094494015154222219432913021739125383559150310033303251117491569691745027149433151558854039221640972291011290355218157628232831823425483261119128009282525619020526301639114772473314857391077758744253876117465786711694147764214411112635835538713610110232679877564102468240322648346417663698066378576813492045302240819727856471983963087815432211669122464159117767322532643356861461865452226812688726844596844241610785401676814208088502800541436131462308210259417375623899420757136275167457318918945628352570441335437585753426986994725470316566139919996826282472706413362221789239031760854289437339356188916512504244040089527198378738648058472689546243882343751788520143956005710481194988423906061369573423155907967034614914344788636041031823507365027785908975782727313050488939890099239135033732508559826558670892426124294736701939077271307068691709264625484232407485503660801360466895118400936686095463250021458529309500009071510582326272932645373821049387249966993394246855164832611341461106802674466373343753407642940266829738652209357016263846485285149036293201991996882851718395366913452224447080459239660281715655156566611113598231122506289058549145097157553900243931535190902107119457300243880176615035270862602537881797519478061013715004489917210022201335013106016391541589578037117792775225978742891917915522417189585361680594741234193398420218745649256443462392531953135103311476394911995072858430658361935369329699289837914941939406085724863968683690326556436421664425760791471086998431573374964883529276932822076294728238153740996154559879825989109371712621828302584811238901196822142945766758071865380650648702613389282299497257453033328389638184394477077940228435988341003583385423897354243956475556840952248445541392394100016207693636846776413017819659379971557468541946334893748439129742391433659360410035234377706588867781139498616478747140793263858738624732889645643598774667638479466504074111825658378878

4548581489629612739984134427260860618724554523606431537101127468097787044640 94
7582803487697589483282412392929605829486191966709189580898332012103184303401 28
4951162035342801441276172885302435598300320420245120728725355811958401491809 692
5339507577840006746552603144616705082768277222353419110263416315714740612385 04
2584598841990761128725805911393568960143166828317632356732541707342081733223 046
2987992804908514094790368878687894930546955703072619009502076433493359106024 54
5086453628935456862958531315337183868265617862273637169577418302398600659148 1
6164049449650117321313895747062088474802365371031150898427992754426853277974 311
3951435741722197597993596852522857452637962896126915723579866205734083757668 73
8842664059909935050008133754324546359675048442352848747014435454195762584735 64
2161981340734685411176688311865448937769795665172796623267148103386439137518 659
4673002443450054499539974237232871249483470604406347160632583064982979551010 95
4183623503030945309733583446283947630477564501500850757894954893139394489921 61
2552559770143685894358587752637962559708167764380012543650237141278346792610 19
9558522471722017772370041780841942394872540680155603599839054898572354674564 23
9058585021671903139526294455439131663134530893906204678438778505423939052473 13
6201294769187497519101147231528932677253391814660730008902776896311481090220 972
4520759167297007850580717186381054967973100167870850694207092232908070383263 45
3452038027860990556900134137182368370991949516489600755049341267876436746384 90
2063964019766685592335654639138363185745698147196210841809618846054560390384 5
5343729141446513474940784884423772175154334260306698831768331001133108690421 93
9031080143784334151370924353013677631084913516156422698475074303297167469640 66
6531527035325467112667522460551199581831963763707617991919203579582007595605 302
3462677579439363074630569010801149427141009391369138107258137813578940055995 00
1835425118417213605572752210352680373572652792241737360575112788721819084490 061
7801388971077082293100279766593583875890939568814856026322439372656247277603 78
9081445883785501970284377936240782505270487581647032458129087839523245323789 60
2984166922548964971560698119218658492677040395648127810217991321741630581055 45
9880130048456299765112124153637451500563507012781592671424134210330156616535 60
2473380784302865525722275304999883701534879300806260180962381516136690334111 138
6538510919367393835229345888322550887064507539473952043968079067086806445096 98
6548801682874343786126453815834280753061845485903798217994599681154419742536 34
4399602902510015888272164745006820704193761584547123183460072629339550548239 55
7137256840232268213012476794522644820910235647752723082081063518899152692889 10
8455571126603965034397896278250016110153235160519655904211844949907789992007 329
4769058685778787209829013529566139788848605097860859570177312981553149516814 67
1769597609942100361835591387778176984587581044662839988060061622984861693533 73
8657877359833616133841338536842119789389001852956919678045544828584837011709 672
1253533875862158231013310387766827211572694951817958975469399264219791552338 57
6623167627547570354699414892904130186386119439196283887054367774322427680913 23
6544948536676800000106526248547305586159899914017076983854831887501429389089 95
0685453076511680333732226517566220752695179144225280816517166776672793035485 15
4204023817460892328391703275425750867655117859395002793389592057668278967764 45
3184040418554010435134838953120132637836928358082719378312654961745997056745 07
1833206503455664403449045362756001125018433560736122276594927839370647842645 67
6338818807565612168960504161139039063960162022153684941092605387688714837989 55

9999112099164646441191856827700457424343402167227644558933012778158686952506949
9364610175685060167145354315814801054588605645501332037586454858403240298717 09
3480910556211671546848477803944756979804263180991756422809873998766973237695 73
7015808068229045992123661689025962730430679316531149401764737693873514093361 83
3216142802149763399189835484875625298752423873077559555955465196394401821840 99
8412489826236737714672260616336432964063357281070788758164043814850188411431 88
5988276944901193212968271588841338694346828590066640806314077577257056307294 0
0492940302420498416565479736705485580445865720227637840466823379852827105784 31
9753541795011347273625774080213476826045022851579795797647467022840999561601 56
9108903845824502679265942055503958792298185264800706837650418365620945554346 13
5134152570065974881916341359556719649654032187271602648593049039787489589066 12
7250794828276938953521753621850796297785146188432719223223810158744450528665 23
8022532843891375273845892384422535472653098171578447834215822327020690287232 33
0053862163479885094695472004795231120150432932266282727632177908840087861480 22
1475376578105819702226309717495072127248479478169572961423658595782090830733 23
3560348465318730293026659645013718375428897557971449924654038681799213893469 24
4741985097334626793321072686870768062639919361965044099542167627840914669856 92
5715074315740793805323925239477557441591845821562518192155233709607483329234 92
1034514626437449805596103307994145347784574699992128599999399612281615219314 88
8769388022281083001986016549416542616968586788372609587745676182507275992950 89
3180521872924610867639958916145855058397274209809097817293239301067663868240 40
1113040247007350857828724627134946368531815469690466968693925472519413992914 652
4238577625500474852954768147954670070503479995888676950161249722820403039954 63
2788306959762493615101024365553522306906129493885990157346610237122354789112 92
5476961760050479749280607212680392269110277722610254414922157650450812067717 35
7120271802429681062037765788371669091094180744878140490755178203856539099104 77
5941413215432844062503018027571696508209642734841469572639788425600845312140 65
9358090412711359200419759851362547961606322887361813673732445060792441176399 759
7461938358457491598809766744709300654634242346063423747466608043170126005205 59
2849369594143408146852981505394717890045183575515412522359059068726487863575 25
4191128887737176637486027660634960353679470269232297186832771739323619200777 45
2212624751869833495151019864269887847171939664976907082521742336566272592844 06
2043021411371992278526998469884770232382384005565551788908766136013047709843 86
1168705231055314916251728373272867600724817298763756981633541507460883866364 06
9347043720668865127568826614973078865701568501691864748854167915459650723428 77
3069985371390430026653078398776385032381821553559732535306860430106757608389 086
2704984188859513809103042359578249514398859011318583584066747237029714978508 41
4585308578133915627076035639076394731145549583226694570249413983163433237897 59
5568085683629725386791327505554254449194358912840504522695381217913191451350 09
9384631177401797151228378546011603595540286440590249646693070776905548102885 020
8085800878115773817191741776017330738554758006056014337743299012728677253043 18
2519757916792969965041460706645712588834697979642931622965520168797300035646 30
4579308840327480771811555330909887025505207680463034608658165394876951960044 08
4820659673794731680864156456505300498816164905788311543454850526600698230931 57
7765003780704661264706021457505793270962047825615247145918965223608396645624 10
5195510522357239739512881816405978591427914816542632892004281609136937773722 29

99833270820829699557377273756676155271139225880552018988762011416800546873655580
63347160373429170390798639652296131280178267971728982293607028806908776866059 3
25274637840539769184808204102194471971386925608416245112398062011318454124478 20
50110798760717155683154078865439041210873032402010685341947230476667217498698
68547076781205124736792479193150856444775379853799732234456122785843296846647 5
13336573692387201464723679427870042503255589926884349592876124007558756946413 7
05625140011797133166207153715436006876477318675587148783989081074295309410605 9
69443158477539700943988394914432353668539209946879645066533985738887866147629 4
43414010498889931600512076781035886116602029611936396821349607501116498327856 35
31614516845769568710900299976984126326650234771672865737857908574664607722834 1
54031144152941880478254387617707904300015669867767957609099669360755949651527 3
63498118964130433116627747123388174060373174397054067031096767657486953587896 70
03192586625941051053358438465602339179674926784476370847497833365557900738419 1
47319886271352595462518160434225372996286326749682405806029642114638643686422 4
72488728343417044157348248183330164056695966886670956349141632842641497453334 9
99948000266998758881593507357815195889900539512085351035726137364034367534714 1
04836017546488300407846416745216737190483109676711344349481926268111073994825 06
07394950735031690197318521195526356325843390998224986240670310768318446607291 2
48747540316179699411397387765899868554170318847788675929026070043212666179192 2
35209382278788809886335991160819235355570464634911320859189796132791319756490 97
60001399623444553501434642686046449586247690943470482932941404111465409239883 44
43515913320107739441118407410768498106634724104823935827401944935665161088463 12
56785297769734684303061462418035852933159734583038455410337010916767763742762 1
02137013548544509263071901147318485749233181672072137279355679528443925481560 9
13728128406333039373562420016045664557414588166052166608738748047243391212955 8
77763906969037078828527753894052460758496231574369171131761347838827194168606 6
25721036851321566478001476752310393578606896111259960281839309548709059073861 35
19145918195102973278755710497290114871718971800469616977700179139196137914171 6
27070189584692143436967629274591099400600849835684252019155937037010110497473 3
94938778858989417433031785348707603221982970579751191440510994235883034546353 49
23498268836240433272674155403016190568065418093940998202060999414021689090070
82133072308966211977553066591881411915778362729274615618571037217247100952142 36
96483086410259288745799932237495519122195190342445230753513380685680735446499 5
12720317448719540397610730806026990625807602029273145525207807991418429063884 4
37349968145827337207266391767020118300464819000241308350884658415214899127610 6
51374153943565721139032857491876909441370209051703148777346165287984823533829 7
26013611098451484182380812054099612527458088109948697221612852489742555551607 6
37167505489617301680961380381191436114399210638005083214098760459930932485102 51
68294467260666138151745712559754953580239983146982203613380828499356705575524 7
12902745397762140493182014658008021566536067765508783804304134310591804606800 8
34591136640834887408005741272586704792258319127415739080914383138456424150940 8
49133918096840251163991936853225557338966953749026620923261318855891580832455 5
71948453875628786128859004106006073746501402627824027346962528217174941582331 7
49239683530136178653673760642166778137739951006589528877427662636841830680190 8
04609849809469763667335662282915132352788806157768278159588669180238940333076 4
41912403412022316368577860357276941541778826435238131905028087018575047046312 9

33353757285386605888904583111450773942935201994321971171642235005644042 97989208
15943071670198574692738486538334361457946341759225738985880016980147574 2054299
58012429581054565108310462972829375841611625325625165724980784920998979 9062003
59365099347215829651741357984910471116607915874369865412223483418877229 29446335
17865385673196255985202607294767407261676714557364981210567771689348491 7660771
70527718760119990814411305864557791052568430481144026193840232247093924 98029335
50731845890355397133088446174107959162511714864874468611247605428673436 70904667
84686702740918810142497111496578177242793470702166882956108777944050484 37528443
37510882826477197854000650970403302186255614733211777117441335028160884 03517814
52541964320309576018694649088681545285621346988355444560249566684366029 221951
24830910605377201980218310103270417838665447181260397190688462370857518 0800353
27047185659499476124248110999288679158969049563947624608424065930948621 5076903
14987020673533848349550836366017848771060809804269247132410009464014373 6032656
45184566792456669551001502298330798496079949882497061723674493612262229 6179081
43114146609412341593593095854079139087208322733549572080757165171876599 4498569
37956238755516175754380917805280294642004472153962807463602113294255916 0025707
35628126387331060058910652457080244749375431841494014821199962764531068 00663118
38237616396631809314446712986155275982014514102756006892975024630401735 1489194
57636078935285550531733141645705049964438909363084387448478396168405184 5273288
40323452024705685164657164771393237755172947951261323982296023945485797 5458651
74587877133181387529598094121742273003522965080891777050682592488223221 5493804
83714547816472139768209633205083056479204820859204754998573203887639160 199524
09189389455767687497308569559580106592650303626615975066222508406742889 826590
75106375635699682115109496697445805472886936310203678232501823237084597 90111548
47208761821247781326633041207621658731297081123075815982124863980721240 7868878
11450165582513617890307086087019897588980745664395515741536319319198107 0575336
63373803827215279884935039748001589051942087971130805123393322190346624 9917169
15094854140187106035460379464337900589095772118080446574396280618671786 1017156
74096766208029576657705129120990794430463289294730615951043090222143937 1849560
63405618934251305726829146578329334052463502892917547087256484260034962 9611654
13823007731332729830500160256724014185152041890701154288579920812198449 3156999
05918201181973350012618772803681248199587707020753240636125931343859554 2547781
96114293516356122349666152261473539967405158499860355295332924575238881 0136202
34762466905581643896786309762736550472434864307121849437348530060638764 4566272
18666170123812771562137974614986132874411771455244470899714452288566294 2440230
18479120547849857452163469644897389206240194351831008828348024924908540 3077863
87516591130287395878709810077271827187452901397283661484214287170553179 6543076
50453432460053636147261818096997693348626407743519992868632383508875668 3595097
26557481543194019557685043724800102041374983187225967738715495839971844 4907279
14196584593008394263702087563539821696205532480321226749891140267852859 9673405
24203109179789990571882194939132075343170798002373659098537552023891164 3467185
58290685371189795262623449248339249634244971465684659124891855629589329 909035
23923333364743520370770101084388003290759834217018554228386161721041760 3011645
91878053936744747205998502358289183369292233732399948043710841965947316 2654825
74809948250999183300697656936715968936449334886474421350084070660883597 235039
53234017958255703601693699098867113210979889707051728075585519126993067 3099250

7040702455685077867906947661262980822516331363995211709845280926303759224267425
7555998928927837047444521893632034894155210445972618838003006776179313813991620
5806270165102445886924764924689192461212531027573139084047000714356136231699237
1694848132554200914530410371354532966206392105479824392125172540132314902740585
8920632175894943454890684639931375709103463327141531622328055229729795380188016
2859073572955416278867649827418616421878988574107164906919185116281528548679417
3638906653885764229158342500673612453849160674137340173572779956341043326883569
5078149317380073623541800706191802673285511919426760912210359874692411728374931
2616339500123959924050845437569850795704622266461900010350049018303415354584283
3764378111988556318777792537201166718539541835984438305203762819440761594106820
7169703022851522505731260930468984234331527321313612165828080752126315477306044
2377475350595228717440266638914881717308643611138906942027908814311944879941715
4042103412190847094080254023932942945493878640230512927119097513536000921971105
4120966831115163287054230284700731206580326264171161659576132723515666625366727
1899853419989523688483099930275741991646384142707798870887422927705389122717248
6322028898425125287217826030500994510824783572905691988555467886079462805371227
0424665431921452817607414824038278358297193010178883456741678113989547504483393
1468963076339665722672704339321674542182455706252479721997866854279897799233957
9057581890622525473582205236424850783407110144980478726691990186438822932305382
3185597328697809222535295910173414073344884761005564018242392192695062083183814
5469839236646136398910121021770959767049083050818547041946643713122996923588953
8493013635657618610606222870559942337163102127845744646398973818855674626087948
2018647487672727222206267646533809980196688368099415907577685263986514625333631
2450536402610569605513183813174261184420189088853196356986962795036738424313011
3317533053298020166888174813429886815855778103432317530647849832106297184251843
8553442762012823457071698853051832617964117857960888815032960229070561447622091
5094739035946646916235396809201394578175891088931992112260073928149169481615273
8427362642980982340632002440244958944561291670495082358124873917996486411334803
2475777521970893277226234948601504665268143987705161531702669692970492831628550
4212898146706195331970269507214378230476875280287354126166391708245925170010714
1808548006369232594620190022780874098597719218051585321473926532515590354102092
8466592529991435379182531454529059841581763705892790690989691116438118780943537
1521332261443625314490127454772695739393481546916311624928873574718824071503995
0094467319543161938554852076657388251396391635767231510055560372633948672082078
0865373494244011579966750736071115935133195919712094896471755302453136477094209
4635696982226673775209945168450643623824211853534887989395673187806606107885440
0055082765703055874485418057788917192078814233511386629296671796434687600770479
9953788338787034871802184243734211227394025571769081960309201824018842705704609
2622564178375265263358324240661253311529423457965569502506810018310900411245379
0153329661569705223792103257069370510908307894799990049993953221536227484766036
1367769797856738658467093667958858378879562594646489137665219958828693380183601
1932368578558558195556042156250883650203322024513762158204618106705195330653060
6065010548871672453779428313388716313955969058320834168984760656071183471362181
2324622725884199028614208728495687963932546428534307530110528571382964370999035
6948885285190402956047346131138263878897551788560424998748316382804046848618938
1895905420398898726506976202019955484126500

0539442820393012748163815853039643992547020167275932857436666164411096256633373
0540921951967514832873480895747777527834422109107311135182804603634719818565557
2957144747682552857863349342858423118749440003229690697758315903858039353352135
8860079600342097547392296733310649395601812237812854584317605561733861126734787
0745850676063048229409653041118306671081893031108871728167519579675347188537229
3096161432040063813224658411111577583585811350185690478153689381377184728147519
9835050478129771859908470762197460588742325699582889253504193795826061621184237
6876851141831606831586799460165205774052942305360178031335726326705479033840127
5730591233960188013782542192709476733719198728738524805742124892118347087662967
6720727232565050565129333126059505777727542471241648312832982072361750574673870
2820957554430596839555568686118839713552208445285264008125202766555767749596968
2661260456524568408613923826576858338469849977872670655519185446869846947849577
3462260629421962455708537127277652309895545019303773216664918257815467729200527
1266714346320963789185232321501897612603437368406719419303774680099929687758247
4104787812326625318184596045385354383911449677531286426092521153767325886672260
4042523491087026958099647595805794663973419064010036361904042033113579336542427
6303561457009011244800890020801478056603710154122328891465722393145076071670648
3556827437743965789067972687438473076346451677562103098604092717090951280863097
0297385044527182892749689212106670081648583395537735919136950153162018908887488
4210798706899114804669270650940762046502772528650728905328548561433160812693007
5693785417861096969202538865034577183176686885923681488475276498468821949739727
9707737187188400414323127636504814531122850990020742409255859252926103021067368
8154347015252348786351643976235860419194129697690405264832347009911154242601273
4380220893310966863678986949779940012601642276092608234930411806438291383473545
6797253992623387915829984864592717340592256207491053085315371829116816372193957
1887009577881815868504645076993439409874335144316263303172477474868979182092397
4808331439708406730840795893581089665647758599055637695252326536144247802308268
8118310377358870892406130313364773710116282146146616794040905186152603600925219
4721889091810733587196414214447865489952858234394705007983038853886083103571937
0600277119455802191194289992272235345870756624692617766317885514435021828702668
5610665003531050216318206017609217984684936863161293727951873078972637353717158
0256378733577180818487845886650433582437700414771041493492743845758710715973
1559439426412570270965125108115548247939403597681188117284721582501094960966257
3933953809221955919181885526780621499231727631632183398969380756168559117529985
4501320671293924041445938623988093812404521914848316462101473891825101090967737
8690664041589736104764365000680771056567184862814963711883219244566394581449147
8616550049567698269030891118568798692947051352481609174324301538368470729289898
2846022237301452655679898627767968091469798378268764311598832109043715611299766
5215396354644208619756737000573876497843768628768179249746943842746525631632378
0055513041742273416464551278127845777724575203865437542828256714128858345444357
1325620544642410110379554641905811686230596447695870540721419852121067343324107
5676757581845699069304604752277016700568454396923404171108988993416350585157878
8735343081552081177207188037910404698306957868547393765643363197978680367187307
7969392423632144845035477631567025539006542311792015346497792906624150832885837
9529054263768766896880503331722780018588506973623240389470047189761934734430847
3744375992503417880797223585913424581314404984770173236169471976571535319775497

π

9716278566311904691260918259124989036765417697990362375528652637573376352696934435440047306719886890196814742876779086697968852250163694985673021752313252926537589641517147955953878427849986645630287883196209983049451987439636907068276265748581043911223261879405994155406327013198989570376110532360629867480377915376751158304320849872092028092975264981256916342500052290887264692528466610466539217148208013050229805263783642695973370705392278915351056888393811324975707133102950443034671598944878684711643832805069250776627450012200352620370946602341464899839025258883014867816219677519458316771876275720050543979441245990077115205154619930509838698254284640725554092740313257163264079293418334214709041254253353232480219322770753555467958716383587501815933871742360615511710131235256334858203651461418700492057043720182617331947157008675785393360786227395581857975872587441025420771054753612940474601000940954449596628814869159038990718659805636171376922272907641977551777201042764969496110562205925024202177042696221549587264539892276976603105249808557594716310758701332088614632664125911486338812202844406941694882615295776253250198703598706743804698219420563812558334364219492322759372212890564209430823525440841108645453694049692714940033197828613181861888111840825786592875742638445005994422956858646048103301538891149948693543603022181094346676400002236255057363129462629609619876056425996394613869233083719626595473923462413459779574852464783798079569319865081597767535055391899115133525229873611277918274854200868953965835942196333150286956119201229888987006079992795411188269023078913107603617634779489432032102773359416908650071932804017163840644987871753756781185321328408216571107549528294974936214608215583205687232185574065161096274874375098092230211609982633033915469494644491004515280925089745074896760324090768983652940657920198315265410658136823791984090645712468948470209357761193139980246813405200394781949866202624008902150166163813538381515037735022966074627952910384068685569070157516624192987244482719429331004854824454580718897633003232525821581280327467962002814762431828622171054352898348208273451680186131719593324711074662228508710666117703465352839577625997744672185715816126411143271794347885990892808486694914139097716736900277758502686646540565950394867841110790116104008572744562938425494167594605487117235946429105850909950214958793112196135908315882620682332156153086833730838173279328196983875087083483880463884784418840031847126974543709373298362402875197920802321878744882872843727378017827008058782410749357514889978911739746129320351081432703251409030487462262942344327571260086642508333187688650756429271605525289544921537651751492196367181049435317858383453862556566440657251363575064353236508936790431702597878177190314867963840828810209461490079715137717099061954969640070867667102330048672631475510537231757114322317411411680622864206388906210192355222354671166213749969326932173704310598722503945657492461697826097025335947502091383667377289443869640002811034402608471289900074680776484408871134135250336787731679770937277868216611786534423173226463784769787514433209534000165069213054647689098505020301504480834261845208730530973189492916425322933612431514306578264070283898409841602950309241897120971601649265613413433422298827909921786042679812457285345801338260995877178113102167340256562744007296834066198480676615805021691833723680399027931606420436812079900316264449146190219458229690992122788553948783538305646864881655562294315673128274390826450611628942803501661336697824051770155219626522725455850738640585299830379180350432876703809252 16

7907571204061237596327685674845079151147313440001832570344920909712435809447906046249431345502890068064870429353403743603262582053579011839564908935434510134
29696175452495739606214902887289327925206965353863964432253883275224996059869747598823299162635459733244451637553343774929289905811757863555556269374269109471170021654117182197505198317871371060510637955585889055688528879890847509157646390746936198815078146852621332524738376511929901561091897779220087057933964638274906806987691681974923656242260871541761004306089043779766785196618914041449252704808819714988015420577870065215940092897776013307568479669929554336561398477380603943688958876460549838714789684828053847017308711177611596635050399793438693391197898871091565417091330826076474063057114110988393880954814378284745288383680794188843426662220704387228874139478010177213922819119923654055163958934474263953824829609036900288359327745855060801317988407162446563997948275783650195514221551339281978226984278638391679715091262410548725700924070045488485692950448110738087996547481568913935380943474556972128919827177020766613602489581468119133614121258783895577357194986317210844398901423948496659251731388171602663261931065366535041473070804414939169363262373767770958503132559900957627319573086480424677012123270205337426670531424482081681303063973787366424836725398374876909806021827857862165127385635132901489035098832706172589325753639939790557291751600976154590447716922658063151110280384360173747421524760851520990161585823125715907334217365762671423904782795872815050956330928026684589376496497702329736413190609827406335310897924642421345837409011693919642504591288134034988106354008875968200544083643865166178805576089568967275315380819420773325979172784376256611843198910250074918290864751497940031607038455494653859460274524474668123146879434416109933389089926384118474252570445725174593257389895651857165759614812660203107976282541655905060424791140169579003383565748692528007430256234194982864679144763227740055294609039401775363356554719310001754300475047191448998410400158679461792416100164547165513370740739502604427695385538343975505488710997852054011751697475813449260794336895437832211724506873442319898788441285420647428097356258070669831069799352606933921356858813912148073547284632277849080870024677763036055512323866562951788537196730346347012229395816067925091532174890308408865160611190114984434123501246469280288059961342835118847154497712784733617662850621697787177438243625657117794500644777183702219991066950216567576440449979407650379999548450027106659878136038023141268369057831904607927652927776940436130230517870805465115424693952651271010529270703066730244471259739399505146284047674316373997825918454117641332790646063658415292701903027601733947486696034869497654175242930640727005059039503148522921392575594845078867977925253931765156416197168443524369794447355964260633391055126826061595726217036698506473281266724521989060549880280782881429796336696744124805982192146339565745722102298677599746738126069367069134081559412016115960190237753525556300606247983261249881288192937343476862689219239777833910733106588256813777172328315329082525092733047850724977139448333892552081175608452966590553940965568541706001179857293813998258319293679100391844099286575605993598910002969864460974714718470101531283762631146774209145574041815908800064943237855839308530828305476076799524357391631221886057549673832243195650655460852881201902363644712703748634421727257879503428486312944916318475347531435041392096108796057730987201352484075057637199253650470908582513936863463863368042891767107602111159828

8755399401200760139470336617937153963061398636554922137415979051190835882900976566473007338793146789131814651093167615758213514248604422924453041131606527009743300884990346754055186406773426035834096086055337473627609356588531097609942383473822220872924644976845605795625167655740884103217313456277358560523582363895320385340248422733716391239732159954408284216666360232965456947035771848734420342277066538373875061692127680157661810954200977083636043611105924091178895403380214265239489296864398089261146354145715351943428507213534530183158756282757338982688985235577992957276452293915674775666760510878876484534936360682780505646228135988858792599409464460417052044700463151379754317371877560398159626475014109066588661621800382669899619655805872086397211769952194667898570117983324406018115756580742841829106151939176300591943144346051540477105700543390001824531177337189558576036071828605063564799790041397618089553636696031621931132502238517916720551806592635180362512145759262383693482226658955769946604919381124866090997981285718234940066155521961122072030922776462009993152442735894887105766238946938894464950939603304543408421024624010487233287500817491798755438793873814398942380117627008371960530943839400637561164585609431295175977139353960743227924892212670458081833137641658182695621058728924477400359470092686626596514220506300785920024882918608397437323538490839643261470005324235406470420894992102504047267810590836440074663800208701266642094571817029467522785400745085523777208905816839184465928294170182882330149715542352359117748186285929676050482038643431087795628929254056389466219482687110428281638939757117577869154301650586029652174595819888786804081103284327398671986213062055598552660364050462821523061545944744899088390819997387474529698107762014871340001225355222466954093152131153379157980269795557105085074738747507580687653764457825244326380461430428892359348529610582693821034980004052484070844035611678171705128133788057056434506161193304244407982603779511985486945591520519600930412710027277849301555038895360338261929343797081874320949914159593396368110627557295278004254863060054523839151068998913578820019411786535682149118528207852130125518518493711503422159542244511900207393539627400208110465530207932867254740543652717595893500716336076321614725815407642053020045340183572338292661915308354095120226329165054426123619197051613839357326693760156914429944943744856809775696303129588719161129294681884936338647392747601226964158848900965717086160598147204467428664208765334799858222090619802173211614230419477754990738738567941189824660913091691772274207233367635032678340586301930193242996397204445179288122854478211953530898910125342975524727635730226281382091807439748671453590778633530160821559911314144205091447293535022230817193663509346865858656314855575862447818620108711889760652969899269328178705576435143382060141077329261063431525337182243385263520217735440715281898137698755157574546939727150488469793619500477720970561793913828989845327426227288647108883270173723258818244658436249580592560338105215606206155713299156084892064340303395262263451454283678698288074251422567451806184149564686111635404971897682154227722479474033571527436819409892050113653400123846714296551867344153741615042563256713430247655125219218035780169240326699541746087592409207004669340396510178134857835694440760470232540755557764728450751826890418293966113301601311190773986324627782190236506603740416067249624901374332172464540974129955705291424382080760983648234659738866913499197840131080155813439791948528304367390124820824448141280954437738983200598649091595053228 5

791457688496257866588599917986752055455809900455646117875524937012455321717019
428288461740273664997847550829422802023290122163010230977215156944642790980219
082668986883426307160920791408519769523555348865774342527753119724743087304361
9511396119080030255878387644206085044730631299277888942729189727169890575925244
679660189704829609491906487646937027507738664323919190422542902353189233773293
166736086996228032557185308919284403805071030064776847863243191000223929785255
372375566213644740096760539439838235764606992465260089090624105904215453927904
411529580345334500256244101006359530039598864466169595626351878060688513723462
707997327233134693971456285542615467650632465676620279245208581347717608521691
3409465203076733918411475041401689241213198268815686645614853802875393311602322
925556189410429953356400957864953409351152664540244187759493169305604486864208
627572011723195264050230997745676476478384889734643172159806267876718380052476968
840849891850861490034324034767426862459523958903585821350064509981782446360873
1775437885967767291952611121385919472545140030118050343787527766440276261894101
757687268042817662386068047788524288743025914524707395054652513533945959878961
977891104189029294381856720507096460626354173294464957661265195349570186001541
262396228641389779673332907056737696215649818450684226369036784955597002607986
799626101903933126376855696876702929537116252800554310078640872893922571451248
113577862766490242516199027747109035593330930494838059785662884478744146984149
906712376478958226329490467981208998485716357108783119184863025450162092980582
920833481363840542172005612198935366937133673339246441612522319694347120641737
5491216357008573694397305979709719726666642267431117762176403068681310351899112
271339724036887000996862922546465006385288620393800504778276912835603372548255
793912985251506829969107754257647488325341412132800626717094009098223529657957
9978030182824284902214707481111240186076134151503875698309186527806588966823625
239378452726345304204188025084423631903833183845505223679923577529291069250432
614469501098610888999146585518818735825281643025209392852580779697376208456374
821144339881627100317031513344023095263519295886806908213558536801610002137408
511544849126858412686958991741491338205784928006982551957402018181056412972508
360703568510553317878408290000415525118657794539633175385320921497205266078312
602819611648580986845875251299974040927976831766399146553861089375879522149717
317281315179329044311218158710235187407572221001237687219447472093493123241070
650806185623725267325407333248757544829675734500193219021991199607979893733836
73242576103938985349278777473980508080015544764061053522203254094435677187945
654304067358964910176107759483645408234861302547184764851895758366743997915085
128580206078205544629917232020282229148869593997299742974711553718589242384938
55858595407438104882646487880533042714630119415898963287926783273224561038521
9701113046658710050008328517731177648973523092666123458887310288351562644602367
1996644554727608310118788389151149340939344750073025855814756190881398752357812
33134227986650352725367171230756861045004548970360079569827626392344107146584
895780241408158405229536937499710665594894459246286619963556350652623405339439
1421112718106910522900246574236041300936918892558657846684612156795542566054160
050712766417660568742742003295771606434486062012398216982717231978268166282499
387149954491373020518436690767235774000539326626227603236597517189259018011042
903842741855078948874388327030632832799630072006980122443651163940869222207453
202446241211558043545420642151215850568961573564143130688834431852808539759277

344336553841883403035178229462537020157821573732655231857635540989540332363823
192198921711774494694036782961859208034038675758341115188241774391450773663840707
188048935825686854201164503135763335550944031923672034865101056104987272647213
198654343545040913185951314518127643731043897250700498198705217627249406521461
995923214231443977654670835171474936798618655279171582408065106379950018429593
879915835017158075983784962257398512129810326379376218322456594236685376799111
314010804313973233544909082491049914332584329882103398469814171575601082970658
306521134707680368069532297199059990445120908727577622535104090239288877942463
048328031913271049547859918019696783532146444118926063152661816744319355081708
187547705080265402529410921826485821385752668815558411319856002213515888721036
569608751506318753300294211868222189377554602722729129050429225978771066787384
000061677215463844129237119352182849982435092089180168557279815642185819119749090
985730570332667646460728757430565372602768982373259745084479649545648030771598
153955827779139373601717422996027353102768719449444917939785144631597314435351
850491413941557329382048542123508173912549749819308714396615132942045919380106
231421774199184060180347949887691051557905554806953878540066453375981862846419
905220452803330626369562649091082762711590385699505124652999606285544383833033277
638599800792922846659503551211245284087516229060262011857775313747949362055496464
010730013488531507354873539056029089335264007132747326219603117734339436733857
591245081493357369116645412817881714540230547506671365182582848980995121391939
956332413365567770980030819102720409971486874181346670060940510214626902804491
596465453301077546954130887141653125448130611924078211886900560277818242350226969
618934435254763357353648561936325441775661398170393063287216690572225974520919
291726219984440964615826945638023950283712168644656178523556516412771282691868
861557271620147493405227694659571219831494338162211400693630743044417328478610
177774383797703723179525543410722344551255558999864618387676490397246116795901
810003509892864120419516355110876320426761297982652942588295114127584126273279090
798807559751851576841264742209479721843309352972665210015662514552994745127631
550917636730259462132930190402837954246323258550301096706922720227074863419005
438302650681214142135057154175057508639907673946335146209082888934938376439399
256900604067311422093312195936202982972351163259386772241477911629572780752395050
562515816031333593823115005186268905306583681299881086632632719806112715488587979
809348791291370749823057592909186293919501472119758606727009254771802575033773
079939713453953264619526999659638565491759045833358579910201271320458390320085
387888163363768518208372788513117522776960978796214237216254521459128183179821
604411131167140691482717098101545778193920231156387195080502467972579249760577272
625913328559726371211201905720771409148645074094926718035815157571514050397610
963846755569298970383547314100223802583468767350129775413279532060971154506484
212185936490997917766874774481882870632315515865032898164228288232746866106592
732197907162384642153489852476216789050260998045266483929542357287343977680495
774091449538391575565485459058976495198513801007958010783759945775299196700547
602252552034453988712538780171960718164078124847847257912407824544361682345239
570689514272269750431873633263011103053423335821609333191218806608268341428910404
151732472160533558499932245487307788229052523242348615315209769384610425828497
149634753418375620030149157032796853018686315724884015266398356895636346574353
217834931998255421173084677452970858395076164582296303244243282377374505170285

606980678895217681981567107816334052667595394249262807569683261074953233905362
230908070814559198373553777487420290390181429373115293346444681512129450975965
343062842153194457271186149000176505581770953024688752632501197052094761594167 6
872778447200019278913725184162285778379228443908430118112149636642465903363419 4
540657183544771912446621259392656620306888520055599121235363718226922531781458
792593750441448933981608657900876165024635197045828895481793756681046474614105
142498870252139936870509372305447734112641354892806841059107716677821238332810
262185587751312721179344448201440425745083063944738363793906283008973306241380
614589414227694747931665717623182472168350678076487573420491557628217583972975
134478990696589532548940335615613167403276472469212505759116251529654568544633
498114317670257295661844775487469378464233737238981920662048511894378868224807 2
793520225017965453437572741639107919729529508129429222053477173041844779156739
917384183117103625243957161527146690058147000026330104526435478659032907332054
683388720787354447626479252976901709120078741837367350877133769776834963442524
199499513883150748775374338494582597655609965559543180409201784971846854973706
962120885243770138537576814166327224126344239821529416453780004925072627651507
890850712659970367087266927643083772296859851691223050374627443108529343052730
788652839773352460174635277032059381791253969156210636376258829375713738407544
064689647831007045806134467312715911946084359358259877828352665311510650416232 9
532904777217408355934972375855213804830509000964667608830154061282430874064559
443185341375522016630581211103345312074508682433943215904359443031243122747138 5
842030390106070940315235556172767994160020393975099897629335325855575624808996
691829864222677502360193257974726742578211119734709402357457222271212526852384 2
958742735015636600931880454933389897415714905441825597380808715652814301026704
602843168192303925352977957658624143927015497408792731310516361191375770089295
648233236482982630246079758757677453771601024908046243018565241617566556001608
591215345562670219268998285537787258314514408265458348440947846317877737479 46
535801699607794055687011923286080411309046293508718271259346687127666948738998 2
459852778649956916546402945893506496433580982476596516514209098675520380830920
323048734270346828875160407154665383461961122301375945157925269674364253192739
003603860823645076269882749761872357547676288995075211480485252795084503395857
083813047693788132112367428131948795022806632017002246033198967197064916374117 5
854851878484012054844672588851401562725019821719066960812627785485964818369621
410721714214986361918774754509650308957099470934337856981674465828267911940611 9
560378453978558392407612763441057667510243075598145527861678159496570625597550
743065210853015979080733437360794328667578905334836695554868039134337201564988
342208933999716414797469386969054800891930671380571715058573071488156499207140
867582596028760564597824237702424698053280566327870419267684671162668794634869
504645074202193739452592626686135529406247813612062026364981999994984051438682
852589563422643287076632993048917234007254717641886853513723326678779217383475
414800228033929973579361524127558295692768372312347989894462743304545667900620
324205163962825884430854383072014956721064605332385372031432421126074244858450
945804940818209276391400085404220235562602185643489941454399504109805918179488
826280520664410863190016885681551692294862030107388971810077092905904807490924
271410189335428184299959881696609938369616443815288772140852680887574882932587
358099056707558170179491619061140019085537448827262009366856044755965574764856

74008177381703307380305476973609786543859382187220583902344443508867499866506040645874346005331827436296177786251808189314436325120510709469081358644051922951293245007883339878842933934243512634336520438581291283434529730865290978330067126179813031679438553572629699874035957045845223085639009891317947594875212639707837594486113945196028675121056163897600888009274611586080020780334159145179707303683519697776607637378533301202412011204698860920933908536577322239241244905153278095095586645947763448226998607481329730263097502881210351772312446509534965369309001863776409409434983731325132186208021480992268550294845466181471555744470966953017769043427203189277060471778452793916047228153437980353967986142437095668322149146543801459382927739339603275404800955223181666738035718393275707714204672383862461780397629237713120958078936384144792980258806552212926209362393063731349664018661951081158347117331202580586672763999276357907806381881306915636627412543125958993611964762610140556350339952314032311381965623632719896183725484533370206256346422395276694356837676136871196292181875457608161705303159072882870071231366630872275491866139577373054606599743781098764980241401124214277366808275139095931340415582626678951084677611866597566016599817808941498575497628438785610026379654317831363402513581416115190209649913354873313111502270068193013592959597164019719605362503355847998096348871803911161281359596856547886832585643789617315976200241962155289629790481982219946226948713746244472909345647002853769495885959160678928249105441251599630078136836749020937491573289627002865682934443134234735123929825916673950342599586897069726733258273590312128874666045146148785034614282776599160809039865257571726308183349444182019353333850712923457743755793440621787113300631060033240539916936826037461766385657588775802012293663532702671006812618251729146082025418928859352444910701382062115538277935652969145765020486432828655579347072096348073726921411868954673227677513356901901537236690368653891612916888878764075254934942497334271811788927599315967193547589880979245252623636590363200708544407845447973482918020820449266706344204375553250505275228337788870408040335319234076856301093477721256390886404131010738178533383160381352808281190408325644018420537467929926220376987180180611226244909092426419858208617511771137890516091403815750033664241560952163281971223350231674226005679412814062172196418427057843289598028823350598282081966662490358577899403331522748177769528436816300885317696947836905806710648280835980466988410981351586549069333195223943632879239905348109878302745001720654336990661177845543646877236318444647680691428280045510746866453928053994091087549391660957316197150331669683099294663491427987808422572206971488755806374803088629951184731871247772919100702275888934869394562895158029653721504096031077612898312635899648934102470360366450586872875890514068412381242473863854279082827338279733268855049358743031602747490631295723497426112215174171531336186224109138695006888358989623492763173164783400774608866555987333382113829928776911495492184192087771606068472874673681886167507221017261103830671787856694812948785048949306308616994879870316051588410828235127415353851336589533294862949449506186851477910580469603906937266267038651290520113781085861618888694795760741358553458515176805197333443349523012039577073962377131603024288720053732099825300897761897312981788194467173116064723147624845755192873278282512718244680782421521646956781929409823892628494376024885227900362021938669648221562809360537317804086372726842669642192994681921490870170753336109479138180406328738759380

4826953558307739576144799727000347288018278528138950321798634521611106660883931
4053226944905455527867894417579202440021450780192099804461382547805858048442441
6404775031536054906591430078158372430123137511562284015838644270890718284481675
7527123846782459534334449622010096071051370608461801187543120725491334999424761
7115633321408934609156561550600317384218701570226103101916603887064661438897773
6318780940711527528174689576401581047016965247557740891644568677171585058326
9943401677202156767724068128366565264122982439465133197359199709403275938550266
9557470231813203243716420586141033606524536939160050644953060161267822648942430
7397166717661231048975031885732165554988342121802846912529086101485527815277620
5623750456375769497734336846015607727035509629049392487088406281067943622418700
4747008368842671022558302403599841645951122485272633632645114017395248086194635
8407837535568856223171155209472230654370926067973510005655493812245754837285450
7117973936157561676416928958052572975223385586113883221711073622658162188424431
7885748879810902665379342666421699091405653643224930133486798815488662866505230
4699723557473842483059042367714327879231642240387776433019260019228477831383760
3253612102533693581262408686669973827597736568222790721583247888864236934639610
6436330873013981421143030600873066616480367898409133592629340230432497492688780
3164360268101130957071614191283068657732353263965367739031766136131596555353849
9939860056515592193675997771793301974468114837110320650369319289452140265091540
6518430993655349333718342529843367991593941746622390038952767381333061774762950
7494386871697845376721949350659087571191772087547710718993796089477451265475750
0187119487073873678589020061737332107569330221632062843206567119209695058576117
3961632326217708945426214609858410237813215817727602222738133495410481003073270
5107799948991977963883530734443457532975914263768405442264784216063122769646960
7156473999043715903323906560726644116438605404838847161912109008701019130726070
1044114143241976796828547885524779476481802959736049439700479596040292746299200
3572099761950140348315380947714601056333446998820822120587281510729182971211910
7876424880354672316916541852256729234429187128163232596965413548589577133208330
9911288775917226115273379010341362085614577992398778325083550730199818459025958
3555989260553299673770491722454935329683300002230181517226575787524058832249085
8212800897479093261007625787704286560069961762121768454789964470506624171021330
3274867962374302291553582007801411653480656474882306150033920689837947662550360
5498228053296628621179306284301704924023019857199789488368971830438051821744190
1476604297524372516834354112170386313794114220952958857980601529387527537990309
3887168357209576071522190027937929278630363726876582268124199338408166021603700
2215471014300737753779269906958712128928801905203160128586182549441335382078480
8346531163265040764242839087012101519423196165226842200371123046430067344206474
7718021353070124098860353399152667923871101706221865883573781210935179775604420
5634694997872511254408545222748109148743072598696020402759411789425812818821590
9523596589791811440776533543217575952555361581280011638467203193465072968079907
9396371496177431211940202129757312516525376801735910155733815377200195244454360
2007184847566341540744232862106099761324348754884743453966598133871746609302050
3507027195298394327142537115576660002578442303107342955153394506048622276496660
8762407932435319929263925373107689213535257232108088981933916866827894828117040
7262450194840970097576092098372409007471797334078814182519584259809624174761010
3825264395513525931188504563626418830033853965243599741693132289471987830842760

00401368074703904097238473945834896186539790594118599310356168436869219485382O
55780395773881360679549900085123259442529724486666766834641402189915944565309
23440650667851948417766779470472041958822043295380326310537494883122180391279
78446100139726753892195119117836587662528083690053249004597410947068772912328
43046353372835199536482743258331191444590178096077828835837301185754365995898
72453192531058811502630754257149394302445393187017992360816661130542625399583
97942971602070338767815033010280120095997252222280801423571094760351925544434
29986767817891045559063015953809761875920358937341978962358931125983902598310
67193304189215109689156225069659119828323455503059081730735195503721665870288
53992138576037035377105178021280129566841984140362872725623214428754302210909
72721073474134975514190737043318276626177275996888826027225247133683353452816
92779591328861381766349857728936900965749562287103024362590772412219094300871
55692625758065709912016659622436080242870024547362036394841255954881727272473
53467783647201918303998717627037515724649922289467932322693619177641614618795
13956699567783068290316589699430767333508234990790624100202506134057344300695
45474682175690441651540636584680463692612742110753990421887161276177870142588
64825775223889184599523376292377915585744549477361295525952226578636462118377
98473700347971408206994145580719080213590732269233100831759510659019121294795
08603640757358750205890208704579670007055262505811420663907459215273309406823
49441590891009220296680523325266198911311842016291631076894084723564366808182
86572196882683584027855007828040434537101836510969517823357430305048526537380
35310741859177056103973950626403554422751561011072617793706347238049906669221
19711942591204450846417463835899382399465173955090008594799901360266742614942
00664671150671754221770387745076735637421547829059110126191575558702389570014
11782264698994491790830179547587676016809410013583761357859135692445564776446
17866711539195135769610486492249008344671548638305447791433009768048687834818
67273375843689272431044740680768527862558516509208826381323362314873333671476
52045087662761495038994950480956046098960432912335834885999029452640028499428
87862403981181488476730121675416110662999553668193123287425702063738352020086
63691311733469731741219153633246745325630871347302792174956227014687325867891
34558379964351358800959350877556356243810493852999007675135513527792412429277
88565888566513247302514710210575352516511814850902750476845518252096331899068
27614435138213662152368890578786699432288160283774820355060160298940091197138
50179871683633744139275973644017007014763706655703504338121113576415018451821
36198234951596010647527125759351853043328755377830575095674254426847122196187
91785607839361445113833356491032564057338986671781239722375193164306170138595
94743678433926709867124522111896908402363274114966012434830989299417380305884
16661307304006758838043211155537944060549772170594282151488616567277124090338
27745629097110134885184374118695655449745736845218066982911045058004299887953
90278043835962824094218605562877884288012755388480372864001944161425749999042
20095952046541705981049899675045119364711727722204361026140797508096869751766
23718774834801612031023468056711264476612374762785219024120256994353471622666
89367521983311181351114650385489502512065577263614547360442685949807439693233
97127377157347099713952291182653485155587137336629120242714302503763269501350
11612952993785864681307226486008270881333538193703682598867893321238327053297
25857382790097826460545598555131836688844628265133798491667839409761353766251

982582496634587719501243840403591408492097337546424744881761840700235695801774
101776969250778148933866725578985645898510568919609243988415692806969833522402
256345704973122452693541938370048431833571965166267215755241934019330990183193
091965829209696562476676836596470195957547393455143374137087615173236772042273
856742791706982045499530959188724349395240944416789988463198455048523936629720
797774528143994182567894577957125524268260899408633173715388962628896294021121
088844273765686245276121303710173007851357154045330415079594477761435974378037
424366469732471384104921243141389035790924160364063140381498314819052517209371
039640268089948325722979545640427017577229041732347960736187878899133183058430
693948259613187138164234672187308451338772190869751049428437693250249816566738
162606159417682525099937416728839517440669325496534031014522253161890092353764
863784828813442098700480962271712264074895719390029185733074601043607291909457
679946149292904279816877294264877299528584346477753869069501489841339245403941
446802636254021186143170312511175776428299146445334089209769616990983726523617̲6
874560589470496817013697490952307208268288789073019001825342580534342170592871
393173799314241085264739094828459641809361413847583113613057610846236683723769
591349261582451622155213487924414504175684806412063652017038633012953277769902
311864802006755690568229501635493199230591424639621702532974757311409422018019̲9
368035026495636955866425906762685687372110339156793839895765565193177883000241
613539562437777840801748819373095020699900890899328088397430367736595524891300
156633294077907139615464534088791510300651321934486673248275907946807879819425
019582622320395131252014109960531260696555404248670549986786923021746989009547
85072567297879476988883109348746442640071818316033165511534276155622405474473
37804924621495213258527698847336269182649174338987824789278468918828054669982
303689939783413747587025805716349413568433929396068192061773331791738208562436
433635359863494496890781064019674074436583667071586924521182997893804077137501
290858646578905771426833582768978554717687184427726120509266486102051535642840
632368481807287940717127966820060727559555904040233178749447346454760628189541
512139162918444297651066947969354016866010055196077687335396511614930937570968
554559381513789569039251014953265628147011998326992200066392875374713135236421
589265126204072887716578358405219646054105435443642166562244565042999010256586
927279142752931172082793937751326106052881235373451068372939893580871243869385
934389175713376300720319760816604464683937725806909237297523486702916910426369
262090199605204121024077648190316014085863558427609537086558164273995349346546
314504040199528537252004957805254656251154109252437991326262713609099402902262
062836752132305065183934057450112099341464918433323646569371725914489324159006
2420206128857329261335968087265000456282845575745965921205303413101118275013069
615098355156320043107846019065654938065425252291619918199596027523277022498557
388248998827074659363557685825605180689642853768507720122203479209939361792682
065901421656159253067379445689490708532635681968318617722682499114726157320358
076462981162440133167378927886892290325933498617970219949819257396176730758344
170985592221701718257127775344915082052784309046194608352174020058386728497094
110232669539214454610662150064106747402070091899119513764669044812672536915371̲6
229079138540393756007783515337416774794210038400230895185099454877903934612222
086506016050035177626483161115332558770507354127924990985937347378708119425305̲5
121436979749914951860535920403830235716352727630874693219622190064260886183676

10334600225547747781364101269190656968649501268837629690723396127628722304114181361006026404403003599698891994582739762411461374480405969706257676472376606554161857469052722923822827518679915698339074767114610302277660602006124687647772881909679161335401988140275799217416767879923160396356949285151363364721954061111717673873725557285229400543617851765023075446938693078734991103521825329297260445532107978877114498988709115112372506042387537348412570860640690520584521227545338480082053024504565176695185769132000428167580549248117805198326460324457928297301291053183856368212062155312886685649565126138922613670640939533345705269869596923503530942245438652786776730275404027022463844835532399147513634410440500923303612714960813554905315390210022995957565837053812619656831442860579566966221547216956208700137277685369608407048333251327931122325071486302069512453950037357233468070946564830892098015348787056334910923660575540508641115214414814346304372327104502776866195310785832333485784029716092521532609255893265560067212435946425506599677177038844539618163287961446081778927217183690888012677820743010642252463480745430047649288555340906218515365435547412547615276977266776977277770583158014121856880117050283652755432148034880044429799980621579045641619572127845089284898064264974270905791290692178072987694779751124473059914605062994689428093103421641662993561482813099887074529271604843363081840412646963792584309418544221635908457614607855856247381493142707826621518554160387020687698046174740080832434366538235455510944949843109349475994467267366535251766270677219418319197719637801570216993367508376005716345464367177672338758864340564487156696432104128259564534984138841289042068204700761559691684303899934836679354254921032811336318472259230555438305820694167562999201337317548912203723034907268106853445403599356182357631283776764063101312533521214199461186935083317658785204711236433122676512996417132521751355326186768194233879036546890800182713528358488844411176123410117991870923650718485785622102110400977699445312179502247957806950653296594038398736990724079767904082679400761872954783596349279390457697366164340535979221928587057495748169669406233427261973351813662606373598257555249650980726012366828360592834185584802695841377255897088378994291054980033111388460340193916612218669605849157148573356828614950001909759112521880039641976216355937574371801148055944229873041819680808564726571354761283162920044988031540210553059707666636274932830891688093235929008178741198573831719261672883491840242972129043496552694272640255964146352591434840067586769035038232057293413298159353304444649682944136732344215838076169483121933311981906109614295220153617029857510559432646146850545268497576480780800922133581137819774927176854507553832876887447459159373116247060109124460982942484128752022446259447763874949199784044682925736096853454984326653686284448936570411181779380644161653122360021491876876946739840751717630751684985635920148689294310594020245796962292456664488196757629434953532638217161339575779076637076456957025973880043841580589433613710655185998760075492418721171488929522173772114608115434498266547987258005667472405112200738345927157572771521858994694811794064446639943237004429114074721818022482583773601734668530074498556471542003612359339731291445859152288740871950870863221883728826282288463184371726190330577714765156414382230679184738603914768310814135827575585364359772165002827780371342286968878734979509603110889919614338666406845069742078770028509367203387232629637856038653216432348815557557018469089074647879122436375566686780676105449550172607 9

114293083128576125448194444947324481909379536900820638463167822506480953181040
657025432760438570350592281891987806586541218429921727372095510324225107971807
783304260908679427342895573555925272380551144043800123904168771644518022649168
1641927401106451622431101700056691121733189423400547959684669804298017362570406
733282129962153684881404102194463424646220745575643960452985313071409084608499
653767803793201899140865814662175319337665970114330608625009829566917638846056
7629729314649114937046244693519840395344491351411936679333019366176636525551491
7498230798707228086085962611266050428929696653565251668888557211227680277274370
891738963977225756489053340103885593112567999151658902501648696142720700591605
616615970245198905183296927893555030393468121976158218398048396056252309146263
844738629603984892438618729850777592879272206855480721049781765328621018747676
689724884113956034948037672703631692100735083407386526168450748249644859742813
493648037242611670426687083192504099761531907685577032742178501000644198412420
739640013960360158381056592841368457411910273642027416372348821452410134771652
9603128408658419787951116511529827814620379139855006399960326591248525308493690
313130100799971913622308660110999291428712493885416120380204113401888872196934
779044975274542880728035093058287544207551348166609278793535665212556201399882
496284787262144323628536765025914504683776352825876521391564809721419296755493
84375582600253168536356731379264758780494459441834291727569883762262618463654
5274349766241113845130548144983631178978448973207671950878415861887969295581973
325069995140260151167552975057543781024223895792578656212843273120220071673057
406928686936393018676595825132649914595026091706934751940897535746401683081179
884645247361895605647942635807056256328118926966302647953595109712765913623318
086692153578860781275991053717140220450618607537486630635059148391646765672320
571451688617079098469593223672494673758309960704258922048155079913275208858378
1117685214269334786921895240622657921043620348852926267984013953216458791151579
05046057971083898337186403802441751134722647254701079479399695355469619726763
255229914654933499663234185951450360980344092212206712567698723427940708857070
474293173329188523896721971353924492426178641188637790962814486917869468177591
7171506691114800207594320120619696377951032270890295660855622254526026104607361
313688690092817210681986185537809820184711541636303262656992834241550236009780
464171085255376127289053350455061356841437758544296779770146602943876872251153
638011917581540281208182556064854107879335989210644272448986189616294134180012
951306836386092941000831366733721530083526962357371753307386533382048421903081
864491840937239440334052449095545580164064607615810103017674884750176619086929
460987692016912021816882910408707095609514704169211470274133900522533408348128
70353030310239196999785974139085936054335996970756044601342424536824960987725813
11024732798520721265724990034682938868723048955622532044636026398542252584164
643242716114198178024825955635449072192265838636626637508359443148776351561457
107455280161596770484427141944351832756984075526779264112617652506159652354571
879566731709133193587616282559207830801852068901515047133403861003100559148178
5211038475454293338918844412051794396997019411269511952656491959418997541839323
464742429070271887522353439367363366320030723274703740712398256202466265197409
01997624520561985576257600087081730832883443818310700545144935458542267857855
19153722923795554943334101744201696000906964156127322977702121795186837635908
2255128816470021992348864043959153018464004714321186360622527011541122283802778

538911098490201342741014121559769965438877197485376431158229838533123071751 1329
619045590079380642766958190148426279912217929479873489018684716765038273285520
590829845298062592503521284519259279865935061329619467962523739725655841578537
445675589980324054921869628884903325608514553443916602262577755129162007727968
526293879375304541810807292858919897153817973434961872329276147478501926114504
132748732429705834084711123337462746172746265824153242710593225062553023147387 5
925172478732288149145591560503633457542423377916037495250249302235148196138116
256391141561032684495807250827343176594405409826976526934457986347970974312449
827193311386387315963636121862349726140955607992062831699942007205481152535339 3
946076850019909886553861433495781650089961649079678142901148387645682174914075
623767618453775144031475411206760160726460556859257799322070337333398916369504
346690694828436629980037414527627716547623825546170883189810868806847853705536
480469350958818025360529740793538676511195079373282083146268960071075175520614 4
337841145499501364324463281933463890509365457145069008644834401804283633905135
781572739733345372842633721740657757710798305175557210367959769018899584941301
959995730179012401939086813565855396619413717944876320798688003716073032205474
235722668968018821234243918859841689722776521940324932273147936692340048489760
590379580946960417542796137825537812239476461478329269765451622902817011004378
460387565441517394339600489153188175766505009516974024156447712936566142539493
688842305174001299205568542898538979426699567770270891465137368922061044154816
621568042198384767308717875902792091759006952734566820265133731115180001814341 2
096260165862982107666352336177400783778342370915264406305407180784335806107296
110555002041513169637304684921335683726540030750982908936461204789111475303704 9
893952833457824082817386441322710002968311940203323456420826473276233830294639
378998375836554559919340866235090967961134004867027123176526663710778725111860 3
540375544874186935197336566217723592293967764632515620234875701137957120962377
234313702120310049651521119760131764190820343734851285260291333491512508311980
285017785571072537314913921570910513096505988599993156086365547740355189816673
353588004821466509974143376118277772335191074121757284159258087259131507460602
563490377726337391446137703802131834744730111303267029691733504770163210661622 7
830027269283365584011791419447808748253360714403296252285775009808599609040936
312635621328162071453406104224112083010008587264252112262480142647519426184325 8
533867538740547434910727100497542811594660171361225904401589916002298278017960
351940800465135347526987776095278399843680869089891978396935321799801391354425
527179102253970108106321430485113782914985113819691430434975001899806816444121 2
327332830719282436240673319655469267785119315277511344646890550424811336143498 4
604849051258345683266441528489713972376040328212660253516693914082049947320486
021627759791771234751097502403078935759937715095021751693555827072533911892334
070223832077585802137174778378778391015234132098489423459613692340497998279304
144463162707214796117456975719681239291913740982925805561955207434243295982898
980529233366415419256367380689494201471241340525072204061794355252555225008748
790086568314542835167750542294803274783044056438581591952666758282929705226127
628711040134801787224801789684052407924360582742467443076721645270313451354167
649668901274786801010295133862698649748212118629040337691568576240699296372493
097201628707200189835423690364149270236961938547372480329855045112089192879829
874467864129159417531675602533435310626745254507114181483239880607297140234725

5207134907983989823552687239509093656678789923837125789762487559904432288953883
37731734894112275707141095979004791930104674075041143538178246463079598955563891884773781341347070246747362112048986226991888517456251732519341352038115863350123913054441910073628447567514161050410973505852762044489190978901984315485280533985777844313933883994310444465669244550885946314081751220331390681596592510546858013133838152176418210433429788826119630443111388796258746090226130900849975430395771243230616906262919403921439740270894777663702488155499322458825979020631257436910946393252806241642476868495455324938017639371615636847859823715902385421265840615367228607131702674740131145261063765383390315921943469817605358380310612887852051546933639241088467632009567089718367490578163085158138161966882222047570437590614338040725853862083565176998426774523195824182683698270160237414938363496629351576854061397342746470899685618170160551104880971554859118617189668025973541705423985135560018720335079060946421271143993196046527424050882225359773481519134385712532585404939460108657937980586201433660788252197178090258173708709164604527279771535099103407364250203863867182205228796944583876529479510486607173902293274554267856697768659399234168341222746630150621553205026553414609952493560508549217565491348309589065361756938176374736441833789742297007035452066631709296075919896277324230902523974438610142630986877339138825186843165010279649114977375828889134503411488659486702154921010843280807834280894172980089832975369406449699031253998639195816014689952208806622854084148642747862819755466292788146216071713818801808405720847158689068369193933818642784545379567192723979723646516675920110579956639625985355127635587681402134098290162968734298507924718460568748283313812591619624761569028759010727331032991406238646083333786382579263023915900035576090324772813388873391780969666014696150317542267511259933155296742133363002229649064809345820081810618021002276645804002782133367585730190113717546727630590443531313190360924890972464279284555499134900051802957070829190525567818899138996251386623193800536113462242946102489540724048571232566288889317221164329478161905548680549434410340906807160880282279596869501336438142682521704728708630101373011552368614169083756757476372397631857570381094433905645644685241830281481079983769185121272019350440418046047216269394457883770901059746932197205581140787759897720720096893822493032368305158626572811146379969831375179376232151112523497343052406221052442343533729056551634066695061658928782187077567941760807129737813351871179316500331555238224877306534441794534153952024244497034101208740721881093882681675120422994049481794494727328947701115741394412284555218284249222406587526891722727806071167540469730080370396187877966948825556146743843925701158295466613586786718976612973112672000729715536130275035561678177654422874421147298816148027052438068176535732755786025058470840132088379328160087690813004924914736825170353822196190390149995234953871059973511434782923394991879366086923013755963685323738067035911442432685615121094042595826393016780171286692392832310576588517140202111969570647998140315056330451415644146231637638099044028162569175764891425697141635984393174332702378123369380430128926263753826677950341693343236075002481757418087503884750949394548962097404854426356371649959499209808842947903636662975260032438563529458447289445471662092974954966168774141208821304770228161164560440072363515811497297392189667373826472047226422212420165601502849713063327958143025160136948255670147809357908896571349261581613469018069650895563101212184918058479

227206918716963163300448580201028606578585912699746376617414639341595695395542
033146280265189511679380745733157598460861737026878676029436777805002446733913
324316698803540732323882818475010516413311895370364884226902704780527424906034
920829547550540034571601840725745369381455311753542107265578356154998744474804
273234578800618731493415660463529797794550753593047956872093167245365472083816
858556060438019770307642460834898761013457093948770029461757920619525492557571
090385251714885252656710453498134198033906415298763436954202560802776144219143
189213939088345431317696851018401038444723489488695209819435319065065553546173
358140455448378847525262539496658699920584176527801253410338964698186424300341
467913806190280596078548880107897055169462152287730901044674624979799926271209
516847795684825833414022664772108433624375937416105367340419547389641978954253
350363018614009515347669614762556518738232924685473569358028960115367917873035
531593783630822486151777705415775765617593585120166929431111388635821596676188 3
032610416465171484697938542262168716140012237821377977413126897726671299202592
201740877007695628347393220108815935628628192856357189338495885060385315817976
067947984087836097596014973342057270460352179060564760328556927627349518220323
614411258418242624771201203577638889597431823282787131460805353357449429762179
678903456816988955351850447832561638070947695169908624710001974880920500952194
363237871976487033922381154036347548862684595615975519376541011501406700122692 7
474393888589943859730245414801061235908036274585288493563251585384383242493252
666087588908318700709100237377106576985056433928543376583425967506537150053 33
514489908293887737352051459333049626531415141386124437935885070944688045486975
358170212908490787347806814366323322819415827345671356443171537967818058195852
464840084032909981943781718177302317003989733050495387356116261023999433259780
126893432605584710278764901070923443884634011735556865903585244919370181041626
208504299258697435817098133894045934471937493877624232409852832762266604942385
129709453245586252103600829286649724174919141988966129558076770979594795306013
119159011773943104209049079424448868513086844493705909026006120649425744710353 5
476578592427081304106185462198818300906345881870387558562749115873754210646679
513464875867715438380185213482819158124625993351601989355951679689328522058247
994210345127158771633452229954188396804488355297533612868372259353900792016669
413390911687588039888288692160023732573615882071635162713328105181876021048521
806755266486739089009079195138058626735124312215691637902277328705410842037841 5
256832887180469879525130732663402785190594173389203585403956770356113293544825
856282876106106982297214209619935093313121711878910787668720445488760894101747
986471378824621539559333332755620094395804345379197822805903959599274369137937
786649409640487778417483364326840262829324062600819080818043909145563519368560
630450891422896452199877988493474777291327972660276584016678901364905087411421
268619698620441269652829810870454798615595453380120115564697997678573892018 62
435993267776894540605082188382279098336271671244900267611784982643770330020818
445900097172352043319947082420987715144497510170556430295428218196700092025156
158441742059336581481349026931115170938722600264586305613256057925609273322655 7
934628080568344392137368840565043430739657406101777937014142461549307074136080
544210029560009566358897789926763051771878194370676149821756418659011616086540
863539151303920131680576903417259645369235080641744656235152392905040947995318
407486215121056183385456617665260639371365880252166622357613220194170137266496

6073252010771947931265282763302413805164907174565964853748354669194523580315301969160480994606814904037819829732360930087135760798621425422096419004367905479049930078372421581954535418371129368658430553842717628035279128821129308351575656599944741788438381565148434229858704245592434693295232821803508333726283791830216591836181554217157448465778420134329982594566884558266171979012180849480332448787258183774805522268151011371745368417870280274452442905474518234674919564188551244421337783521423865979925988203287085109338386829906571994614906290257427686038850511032638544540419184958866538545040571323629681069146814847869659166861842756798460041868762298055562963045953227923051616721591968675849523635298935788507746081537321454642984792310511676357749494622952569497660359473962430995343310404994209677883827002714478494069030707324910644415169605325656058677875741747211082743577431519406075798356362914332639781221894628744779811980722564671466405485013100965678631488009030374933887536418316513498254669467331611812336485439764932502617954935720430540218297487125110740401161140589991109306249232128131163405492625713567218186289327861388337180285350565035919527414008695109261675414767926680321092374670872136062783329223864136195941213392780361182763241060047409711110481400036233427145144833346416754663546997314947566434236594934968458845515241507563766050866328274247941360628760412906449138285194564026431532258586240431418386695906332450630003922131926476259626915109044576953014440546180378575030366862124622786397527466678701210033929848733750144756003221006223580293437749550320370127384681630610265703008722754629667968808905871276763610662257223522297392064430935243272281008599730951325286306011054979156447918450046180467624089289256809129305929606423570210615246462050232489665939873249339673769520239917608984745718435319366465291258480644801965201628387951894993367592414856261369959453072872545324632915291101287637706557060953137752775186792329213495524513308986796916512907384130216757323863757582008036357572800275449032795307990079944254110872569318801466793559583467643286887696661009739574996783659339784634695994895061049038364740950469522606385804675807306991229047408987916687211714752764471160440195271816950828973353714853092893704638442089329977112585684084660833993404568902678751600877546126798801546585652206121095349079670736553970257619943137663996060606110640695933082817187642604357342536175694378484849525010826648839515970040590598380812105221111091943323951136051446459834210799058082093716464523127704023160072138543723461267260997870385657091998507595634613248460188409850194287687902268734556500519121546544063829253851276317663922050938345204300773017029940362615434001322763910912988327863920412300445551684054889809080779174636092439334912641164240093880746356607262336695842764583698268734815881961058571835767462009650520606592926354829149904576830721089324585707370166071739819448502884260396366074603118478622583105658087087030556759586134170074540296568763477417643105175103673286924555858208237203860178173940517513043799468882230044378043103170921034261674998000073016094814586374488778522273076330495383944345382770608760763542098445008306247630253572781032783461766970544287155315340016497076657195985041748199087201490875686037783591994719343352772947285537925787684832301101859365800172911869676176550537750302930338307064489128114120255061508964110076238245744886551825810581403453201247547232690875475070785776597325428444593530449920700145387489482265564422236963655441942254413382122254774975354946248276805333369832841561386923

634433585538684711114304982483989918031654586382893537991305352228334301379533 7

295401625762322808113849949187614414132293376710656349252881452823950620902235

787668465011666009738275366040544694165342223905210831458584703552935221992827

276057482126606529138553034554974455147034493948686342945965843102419078592368

022456076393678416627051855517870290407355730462063969245330779578224594971042

018804300018388142900817303945050734278701312446686009277858181104091151172937 4

873627887874907465285556543474888683106411005102302087510776891878152562273525 1

550379532444857787277617001964853703555167655209119339343762866284619844026295

252183678522367475108809781507098978413086245881522660963551401874495836926917

799047120726494905737264286005211403581231076006699518536124862746756375896225

299116496066876508261734178484789337295056739007878617925351440621045366250640

463728815698232317500596261080921955211150859302955654967538862612972339914628 3

584760486276270273097392020014322487075823373549152460856082103288829741839064

788699232736913600488374366152235170584377055452108155133612621429118156153017

588825735948925071088792621286413924433093837973338678061317952373152667738208

580247014335270092438032669517421195076708843263464427491275589077468635821621

660427413151702124585860562336314931646469139465624974717419583542186077487110

573384584336899396459137406033821593522435947516262391886853078228217639832373

061802042465604752794310479618972429953302979249748168405289379104494700459 08

649918727273454135081019838818646736093925719305119686456018557824502182310658

894379865224320506773799661969554724405859224179530068204517953700434724517628

935667705084902131077366257516973355274623029430312035962609534235743972496592

110106578178261087453188748031874308235736991951563409571627009924449297491054

898515196586647401482251063353679497371425102293418825851173719944991150975837 4

613010550506419772153192935487537119163026203032858865852848019350922587577559

742527658401172134232364808402714335636754204637518255252494432965704386138786

590196573880286840189408767281671413703366173265012057865391578070308871426151

907500149257611292767519309672845397116021360630309054224396632067432358279788 9

332324405779199278484633339777737655901870574806828678347965624146102899508487

399692970750432753029972872297327934442988646412725348160603779707298299173029

296308695801996312413304939350493325412355071054461182591141116454534710329881 0

478440677801380771314654000993863064812666143308582068113958383191695455582594

268957698414288937434670841079463189325391069639557807060212459748982935646135

607889834724199794785643620420946134123876131988653523583129968622689486084084

566556068769545012744866314050547353517468730098063227804689122468214608067276

277084024022661554850240089528916571176174390203375848778429112896232470591918 7

469104200584832614067733375102719565399469716251724831223063391932870798380074

848572651612343493327335666447335855643023528088392434827876088616494328939916

639921048830784777704804572849145630335326507002958890626591549850940797276756

712979501009822947622896189159144152003228387877348513097908101912926722710377

889805396415636236416915498576840839846886168437540706512103906250612810766379

904790887967477806973847317047525344215639038720123880632368803701794930895490

077633152306354837425681665336160664198003018828712376748189833024683637148830

925928337590227894258806008728603885916884973069394802051122176635913825152427

867009440694235512020156837777885182467002565170850924962374772681369428435006

293881442998790530105621737545918267997321773502936892806521002539626880749809

26434580116557158867004435039765053234782873273688408635400027406767838219635222265392909398073673913640828987220177767471681181958561337215831190546829360832369761134502817578302029348459829250008956826302712632958662921476531422333517930933879513570953463771836840924444220963193312956203055755173400679737406141621079236334238056468500920371671526425563718538895714164197723874226105966673969971731681694154350952831935564177056686222152179911513556397071433128936575538446483262012064243380169558626985610224606460693307938478588143674070005997697036490192733288261353293631124036506986521606389872502672380874033967443978302582968942568967418643361349794752455262914265228424192430833881035800537870239995421721136865502753413622116931406946695131869281025747959856051450050217159133177516099578655519818861932112821107094422872404424811534060558959583558152320121846058205635926993034788511320686266275887714460359966561084307256965005630644891875994665967728471715395736121081808415472731426617489331341746326623542220720712600146012701206934639520564445543291662986660783089068118790090815295063626782075614388157813511346953663038784120923469428687308393204323338727754968052103028215443247233888452153437272501285897476914608083144041258681815400491877228786980185345453700652665564917091542952275670922221747411206272065662299898060328916720687436549482461086973672255474048128892424718543236057534116728507575520571311566979545848873987422281358879858407831350605482905514827852948911219053831956242287194847594078593980479010941940706717664390327307121358873850499936388382055016834027774960702768448802819122206368886368110435695293006521955282615269912716372773884189932871305634646882273982887631986457098363089177864870866761854856800476725526754147428510281458074031529921978145577568436811101853174981670164266478840902626882444825802753209454991510451851771654631180490456798571325752811791365627815811128881656228587603087597496384943527567661216895926148503078536204527450775295063101248034180458405943292607985443562009370809182152392037179067812199228049606973823874331262673030679594396095495718957721791559730058869364684557667609245090608820221223571925453671519183487258742391941089044411595993276004450655620646116465566548759424736925233695599303035509581762617623184956190649483967300203776387436934399982943020914707361894793269276244518656023955905370512897816345542332011497599489627842432748378803270141867695262118097500640514975588965029300486760520801049153788541390942453169171998762894127722112946456829486028149318156024967788794981377721622935943781100444806079672429276249510784153446429150842764520002042769470698041775832209097020291657347251582904630910359037842977572651720877244740952267166306005469716387943171196873484688738186656751279298575016363411314627530499019135646823804329970695770150789337728658035712790913767420805655493624465

Afterword

One of the intriguing aspects of mathematics is the unending relationships that one finds in what, on the surface, would appear to be disconnected branches of the subject. From my early days I was always fascinated with the ubiquity of π. We know from our school days that π is defined to be the ratio of the circumference of a circle to its diameter. Yet π seems to pop up almost everywhere in mathematics—even outside of the field of geometry. This lovely book took you through a very broad spectrum of the history and appearances of π. Implicit in this definition is the assumption that π has the same value for all circles, whether large or small. This property of the circle has been known for so long that it appears now to be all but impossible to give an account of the time, place, and circumstance of its discovery.

For me the things that stand out most about π are those that have shown the true genius of some of our greatest mathematicians. Specifically, Archimedes (287–212 BCE) and Leonhard Euler (1707–1783) showed some of their greatest brilliance when it came to working with π. Earlier in this book you were exposed to their work. Now, I would like to recap the brilliant insight that these mathematicians exhibited.

Archimedes' attempt to determine the value of π was based on his assumption that the circle's circumference lies between the perimeters of an inscribed and circumscribed regular polygon of the same number of sides, and, as the number of sides increases without limit, these perimeters approach arbitrarily close to the circumference. We start with a circle having unit radius. Hence the circumference is equal to 2π, and the semicircumference is equal to π. We inscribe in this circle an equilateral triangle having semiperimeter b_1 and circumscribe an equilateral triangle having semiperimeter a_1 (fig. 1).

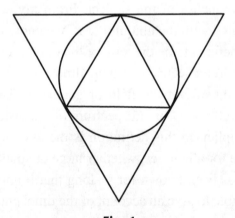

Fig. 1
A circle with unit radius together with inscribed and circumscribed equilateral triangles

Clearly b_1 and a_1 are very crude approximations to π; b_1 is less than π and a_1 is greater than π, or in symbols

$$b_1 < \pi < a_1 \tag{1}$$

To get a better approximation to π, we double the number of sides of the inscribed and circumscribed regular polygons. Thus, we inscribe in the circle a regular hexagon having semiperimeter b_2 and circumscribe a regular hexagon having semiperimeter a_2 (fig. 2). As before, b_2 and a_2 are clearly approximations to π, still crude but better than before; b_2 is less than π and a_2 is greater than π:

$$b_2 < \pi < a_2 \tag{2}$$

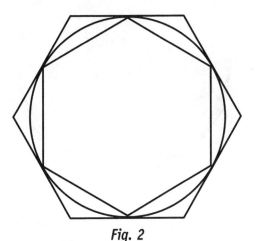

Fig. 2
The circle with unit radius together with inscribed and circumscribed regular hexagons

Furthermore, as comparison of figures 1 and 2 shows, it is obvious too that

$$\pi > b_2 > b_1 \text{ and } \pi < a_2 < a_1 \tag{3}$$

We continue in this way, as Archimedes did more than two thousand years ago, doubling the number of sides, called N, in both inscribed and circumscribed regular polygons until we arrive at $N = 96$ (see fig. 3 for the case that $N = 12$):

$$N = 3, 6, 12, 24, 48, 96 \tag{4}$$

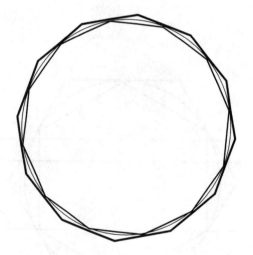

Fig. 3
A circle with unit radius together with inscribed and circumscribed dodecagons (12-gons)

The corresponding semiperimeters b_1, b_2, ..., b_6 of the six inscribed regular polygons and semiperimeters a_1, a_2, ..., a_6 of the six circumscribed regular polygons yield better and better approximations for π, as is already obvious when $N = 12$ (fig. 3).

As before

$$b_n < \pi < a_n, \, n = 1, 2, \ldots, 6 \tag{5}$$
$$\pi > b_6 > b_5 > \ldots b_1, \text{ and } \pi < a_6 < a_5 < \ldots < a_1 \tag{6}$$

It is now easily verified that, owing to the successive doubling of the number of sides in the inscribed and circumscribed regular polygons, N is given by the simple formula

$$N = 3 \cdot 2^{n-1}, \, n = 1, 2, \ldots, 6 \tag{7}$$

so that when $n = 1$, $N = 3$; when $n = 2$, $N = 6$; when $n = 3$, $N = 12$; ...; until finally, when $n = 6$, $N = 96$.

Next, we refer to Archimedes who showed how to calculate the values of a_n and b_n recursively by means of a pair of remarkable formulas:

$$\frac{1}{a_{n+1}} = \frac{1}{2}\left(\frac{1}{a_n} + \frac{1}{b_n}\right) \tag{8}$$

$$b_{n+1} = \sqrt{a_{n+1} b_n} \tag{9}$$

One starts with the values of a_1 and b_1, readily obtained by elementary geometry:

$$a_1 = 3\sqrt{3} \approx 5.196152 \text{ and } b_1 = \frac{3}{2}\sqrt{3} \approx 2.598076 \tag{10}$$

then sets $n = 1$ in equation (8) to calculate first the value of a_2 and then, using equation (9), again with $n = 1$, to calculate the value of

b_2. Then, setting $n = 2$ in equation (8) and using the known values of a_2 and b_2, one calculates the value of a_3. Setting $n = 2$ in equation (9) and using the values of a_3 and b_2, now known, one then calculates the value of b_3, and so on. Archimedes proceeded in this way to find the values of $a_1, a_2, \ldots, a_6$ and $b_1, b_2, \ldots b_6$. Although decimal notation was not known to Archimedes, we may use this notation to briefly summarize his result:

$$a_6 \approx 3.1426 \text{ and } b_6 \approx 3.1410, \tag{11}$$

which yield the value of π correct to two decimal places and, finally,

$$\frac{223}{71} \approx 3.1408 < b_6 \approx 3.1410 < \pi \approx 3.14159 < 3.1426 \approx a_6 < 3.1429 \approx \frac{22}{7} \tag{12}$$

thus arriving at his famous estimate

$$\frac{223}{71} < \pi < \frac{22}{7} \tag{13}$$

Now Archimedes stopped at this point probably because the calculations were becoming too arduous, but there is no need for us to stop since both the decimal system and hand calculators are now available. After a brief calculation we find, for example,

$$\pi < a_{13} \approx 3.1415\ 9272 \text{ and } \pi > b_{13} \approx 3.1415\ 9262 \tag{14}$$
$$\pi < a_{14} \approx 3.1415\ 9267 \text{ and } \pi > b_{14} \approx 3.1415\ 9265 \tag{15}$$

which yield the value of π correct to seven decimal places.

Of course, with the availability of high-speed automatic computers nowadays, we can go much further. In particular, we can find in a matter of seconds the values of $a_{30}, a_{40}, b_{30}, b_{40}$, and so on, yielding the value of π correct to some thirty decimal places at least.

We cannot leave the remarkable formulas in equations (8) and (9) without further comment.

Notice first that $\frac{1}{a_{n+1}}$, as given by equation (8), is the arithmetic mean of $\frac{1}{a_n}$ and $\frac{1}{b_n}$; its value therefore lies between the values of $\frac{1}{a_n}$ and $\frac{1}{b_n}$.

Alternatively, a_{n+1} is said to be the harmonic mean of a_n and b_n, and, again, its value lies between the values of a_n and b_n. Hence, since with increasing n, a_n and b_n approach closer and closer to π, a_n from above and b_n from below, we naturally anticipate that a_{n+1} will be closer to π than either a_n or b_n, as the geometric interpretation already suggested.

In a similar way, b_{n+1}, as given by equation (9), is said to be the geometric mean of a_{n+1} and b_n; its value lies between the values of a_{n+1} and b_n, and, as before, one naturally anticipates that b_{n+1} will be closer to π than either a_{n+1} or b_n. Comparison of the known value of $\pi = 3.14159265358979323846264338\ldots$ with the values of a_n and b_n confirms these expectations.

In this connection one should consult figure 4 and observe that the chord $\overline{AB}$ is a better approximation to the length of the circular arc $\overparen{AB}$ than the sum of the lengths of $\overline{AC}$ and $\overline{BC}$; in symbols

$$m\,\overparen{AB} - AB < AC + BC - m\,\overparen{AB} \qquad (16)$$

Replacing $AC + BC$ by CC', we find

$$m\,\overparen{AB} - AB < CC' - m\,\overparen{AB} \qquad (17)$$

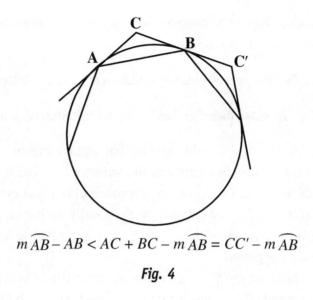

$$m \, \overset{\frown}{AB} - AB < AC + BC - m \, \overset{\frown}{AB} = CC' - m \, \overset{\frown}{AB}$$

Fig. 4

and infer that

$$\pi - b_n < a_n - \pi \tag{18}$$

or that b_n is a better approximator to π than is a_n, as is already suggested, for example, by equation (12).

Of course there is nothing sacred about the regular polygons having sides $N = 3, 6, 12, 24,\ldots.$ One could just as well start with the inscribed and circumscribed square, doubling the number of sides in succession, thus leading to the sequence

$$N = 4, 8, 16, 32, 64, \ldots \tag{19}$$

corresponding to

$$n = 1, 2, 3, 4, \ldots \tag{20}$$

respectively, so that

$$N = 2^{n+1} \tag{21}$$

Now denoting by a_1 and b_1 the semiperimeters of the circumscribed and inscribed squares, respectively, one finds

$$a_1 = 4 \text{ and } b_1 = 2\sqrt{2} \approx 2.82842712 \tag{22}$$

instead of equation (10).

If one now defines a_n and b_n to be the semiperimeters of the circumscribed and inscribed regular polygons having $N = 2^{n+1}$ sides, respectively, one can still calculate a_n and b_n using the same recursion formula, equations (8) and (9), as before. Since the starting point, equation (22), is now different from equation (10), we obtain different sequences

$$a_1, a_2, a_3, \ldots; b_1, b_2, b_3, \ldots \tag{23}$$

which however approach closer and closer to π. The reader may wish to carry out this calculation remembering now to start with equation (22) [not equation (10)] and to use the same recursion formulas, equations (8) and (9), as before to calculate a_n and b_n when $n > 1$.

It is impossible to leave Archimedes without at least brief mention of two of his greatest achievements, especially since they provide our first illustrations of the ubiquitous nature of π. They are nothing less than his famous formulas expressing the values of the volume and surface area of a sphere in terms of its radius, formulas of which Archimedes himself was, for good reason, particularly proud. If one denotes by V and S the volume and surface area, respectively, of the sphere with radius r, Archimedes found that $V = \frac{4}{3}\pi r^3$ and $S = 4\pi r^2$, which are easy consequences of Archimedes' analysis of the sphere and the circumscribed cylinder in which he showed that the volume and surface area of the sphere are equal, respectively, to two-thirds of the volume and surface area (including the bases) of the circumscribed cylinder. According to the historian Plutarch, Archimedes himself expressed the wish during his lifetime that there should be placed on his tomb a sphere together with circumscribed cylinder and an inscription giving the ratio between the volumes of these two bodies that he had discovered.

The subject of infinite series has fascinated mathematicians for centuries. Of particular interest were questions of convergence, and, when convergent, questions concerned with their sums were paramount. It was not until the seventeenth century however that interest in the summation of infinite series became particularly intense, and, largely under the influence of the mathematician Jakob Bernoulli, great progress was made. There was however one particular series the summation of which presented an insuperable obstacle. This series was simply the sum of the reciprocals of the squares of all the integers:

$$1 + \frac{1}{2^2} + \frac{1}{3^2} + \frac{1}{4^2} + \cdots \tag{24}$$

a series long known to be convergent, but the sum of which, despite its apparent simplicity, had resisted all attempts at solution. Even Bernoulli, who had succeeded in summing far more complex appearing series than this one, finally had to admit defeat, but not before he had challenged the mathematical community to solve the so-called Basel problem, the summation of the series (24). The difficulty of the problem was such, however, that it was not until 1735, decades after Bernoulli's death, that the young Leonhard Euler, at the age of twenty-eight, produced the solution:

$$1 + \frac{1}{4} + \frac{1}{9} + \frac{1}{16} + \cdots = \frac{\pi^2}{6} \tag{25}$$

Can anything be more beautiful and unexpected? After all, who would have anticipated that the number π, in view of its definition, should be so intimately related to the integers?

Euler's solution was a model of simplicity and clarity. He simply expressed the function $\frac{\sin x}{x}$ in two ways, first as an infinite series and, second, as an infinite product.

Thus

$$\frac{\sin x}{x} = 1 - \frac{x^2}{3!} + \frac{x^4}{5!} - \frac{x^6}{7!} + \cdots = \left(1 - \frac{x^2}{\pi^2}\right)\left(1 - \frac{x^2}{4\pi^2}\right)\left(1 - \frac{x^2}{9\pi^2}\right)\cdots \tag{26}$$

Comparing these two expressions yields, after a straightforward analysis, the sum in equation (25). Not only did this argument yield the sum in equation (25) but also a very profound extension that enabled Euler to sum the reciprocals of the fourth powers, the reciprocals of the sixth powers, and so on. Thus

$$1 + \frac{1}{2^4} + \frac{1}{3^4} + \frac{1}{4^4} + \cdots = \frac{\pi^4}{90} \tag{27}$$

$$1 + \frac{1}{2^6} + \frac{1}{3^6} + \frac{1}{4^6} + \cdots = \frac{\pi^6}{945} \tag{28}$$

and so on. Equations (25)–(28) naturally cause one to wonder why π, the ratio of the circumference to the diameter of a circle, should be so unexpectedly related to the even powers of the integers.

Euler's derivation of equations (25)–(28) naturally raised the question of the summability of the reciprocals of the odd powers of the integers such as, for example,

$$\frac{1}{1^3} + \frac{1}{2^3} + \frac{1}{3^3} + \cdots \tag{29}$$

$$\frac{1}{1^5} + \frac{1}{2^5} + \frac{1}{3^5} + \cdots \tag{30}$$

and so on. Despite the most intense efforts of many mathematicians since Euler's time to sum these series, none have succeeded. Except for Roger Apéry's[1] interesting result that the series (29) represents an irrational number, virtually nothing is known about the other series in this family.

It is therefore not only a matter of some surprise but of great interest that the alternating series of the odd powers of the odd integers have yielded their secrets. Thus, for example,

$$1 - \frac{1}{3} + \frac{1}{5} - \frac{1}{7} + \cdots = \frac{\pi}{4} \tag{31}$$

$$1 - \frac{1}{3^3} + \frac{1}{5^3} - \frac{1}{7^3} + \cdots = \frac{\pi^3}{32} \tag{32}$$

$$1 - \frac{1}{3^5} + \frac{1}{5^5} - \frac{1}{7^5} + \cdots = \frac{5\pi^5}{1,536} \tag{33}$$

and so on. Each of these sums is seen to be the product of a rational number and an odd power of π reminiscent of the series (25), (27), and (28), where the even powers of π were involved. It is astonishing to find that π is so simply related in some mysterious way to the integers.

As you saw in this book, there are many curiosities attached to π, each fascinating in its own way. Consider the following: a number is said to be square-free if none of its divisors other than unity is a perfect square. For example, the number 15 is square-free since its only divisors greater than unity are 3, 5, and 15, none of which is a perfect

---

1. Roger Apéry was a French mathematician (1916–1994).

square. The number 45 on the other hand is not square-free since it is divisible by $9 = 3^2$, a perfect square.

What is the probability that a number chosen at random be square-free? Would you believe that the answer is $\frac{6}{\pi^2} \approx 0.6079$? If you have difficulty accepting this, I would suggest that you put it to the test. Choose one hundred numbers at random and count the number of them, say m, which are square-free. Is the ratio $\frac{m}{100}$ approximately equal to 0.6079?

Better still, count all the numbers less than or equal to 100 that are square-free. Do you get 61? Is not the ratio $\frac{61}{100} = 0.61$ approximately equal to $\frac{6}{\pi^2} = 0.6079$?

Alternatively, setting $\frac{6}{\pi^2} \approx 0.61$, one finds

$$\pi \approx \left(\frac{6}{0.61} \right)^{\frac{1}{2}} \approx 3.136 \tag{34}$$

not a bad approximation to π, obtained purely experimentally.

Of course, to choose all numbers less than 100 is not the same thing as choosing one hundred numbers at random. Hence, strictly speaking, further justification is needed to validate this procedure, which is beyond the scope of this afterword.

If you are not happy with the rather crude approximation to π obtained with a sample of one hundred, you are free to do the same experiment with a larger sample size, say, one thousand numbers. Is this a good way to estimate the value of π?

Another curiosity, again far afield from the foundation of the concept of π, geometry, is one involving relatively prime numbers. Two numbers are said to be relatively prime if they have no common divisor other than unity. For example, the numbers 10 and 21 are relatively prime since they have no common divisor greater than one. The pair of numbers 15 and 24, on the other hand, are not relatively prime since 3 divides both of them.

What is the probability that two numbers, p and q, chosen at random be relatively prime? Incredibly, the answer is, once again, $\frac{6}{\pi^2}$. As in the previous section, this result may be used to estimate the value of π experimentally. Convince yourself!

These curiosities provide further convincing evidence that the number π occurs frequently, in many diverse contexts. We have seen it appear as the ratio of the circumference of a circle to its diameter (by definition); in formulas that measure the volume and surface area of a sphere; as the sum, in many ways, of infinite series; and, finally, as a measure of probabilities. Can one hope to find more compelling examples that demonstrate the central importance of the number π as well as the interrelatedness of all of mathematics?

I must admit it was my pleasure to be asked to write this afterword since it again brought me into discussions on mathematics with Dr. Alfred S. Posamentier. Gradually, my enthusiasm for a fresh look at π grew. I revisited Archimedes' work and this time discovered things I hadn't seen before—resulting in even greater amazement at this brilliant mathematician. With the supercomputer I was able to dabble into areas not conceivable in my youth. For example, I discovered this absolutely gorgeous gem, which I cannot resist in sharing with the readership:

$$1 - \frac{1}{3^{11}} + \frac{1}{5^{11}} - \frac{1}{7^{11}} + \cdots = \frac{19 \cdot 2,659}{2^{10} \cdot 3^4 \cdot 5^2 \cdot 7} \pi^{11}$$

Who knows where this ubiquitous number will come up next? There were even times during my Nobel Prize–winning research[2] in crystallography that π would often appear. This delightful book guided you through a clear discussion of what π actually represents,

---

2. Dr. Hauptman won the Nobel Prize for Chemistry in 1985, and was lauded as the first mathematician to win a Nobel Prize.

where it appears, how it can be used, its many curious properties, and the history that led us to its known value today. Just as I hadn't taken a new look at π until I was asked to write this afterword— only to become further enchanted with this incredible number— this book gave you, as well, the opportunity to revisit some elementary mathematics with the entertainment provided by this fascinating number called π. I suspect your amazement grew with each section of the book.

Dr. Herbert A. Hauptman
April 2004
Nobel Laureate (Chemistry 1985)
President of the
Hauptman-Woodward Medical Research Institute
Buffalo, New York

Appendixes

Appendixes

Appendix A

A Three-Dimensional Example of a Rectilinear Equivalent to a Circular Measurement

When it comes to circle measurements, π always plays an important role. It rarely, if ever, plays a role in the measurement of a rectilinear figure (one comprised of straight lines). Consequently, it is rare that the measurement of a circular figure is exactly equal to that of a rectilinear figure (an exception in the plane can be found on p. 46). We shall now show an example in three dimensions, in which a circular and a rectilinear figure are equal in volume.

A famous theorem in geometry was developed by the Italian mathematician (Francesco) Bonaventura Cavalieri (1598–1647)[1] and is known today as Cavalieri's principle. It states that "two solid figures are equal in volume if a randomly selected plane cuts both figures in equal areas." The well-known mathematics historian Howard Eves developed a clever proof that "there exists a tetrahedron which has the same volume as a given sphere," or, as he says, where the two solids

1. A student of Galileo.

are "Cavalieri congruent." For this effort, he won the 1992 George Polya Award.[2] The beauty of this discovery is the profoundness of the statement and the relative simplicity of the proof—it is a result that, as the award statement says, would have had "geometers of ancient times inscribe it on their tombstones."

Let us now look at this clever proof. Notice the unusual role that π plays in the proof.

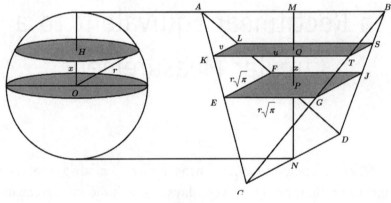

Fig. A-1

We begin with a sphere and two parallel planes, tangent at each of its "north and south" poles. Next we draw two line segments, $\overline{AB}$ and $\overline{CD}$, of length $2r\sqrt{\pi}$ in each of the planes, respectively, but in such a way that they are oriented in perpendicular directions and that the line joining their midpoints is a common perpendicular. We now join the endpoints of these two segments to form the tetrahedron $ABCD$.

We now pass two more planes parallel to the first two: one through the center of the sphere and one x units above it. The former plane cuts the tetrahedron in a square whose side has length $r\sqrt{\pi}$. The reason for this is that the segment $\overline{EF}$ is one joining the

2. Howard Eves, "Two Surprising Theorems on Cavalieri Congruences," *College Mathematics Journal* 22, no. 2 (March 1991): 123–24.

midpoints of two sides of a triangle and is therefore half the length of the third side. The plane x units above the plane through the center of the sphere and parallel to it will cut the tetrahedron in a rectangle with sides of length u and v ($\overline{KT}$ and $\overline{KL}$).

The circle on which the noncenter plane cuts the sphere has a radius of $\sqrt{r^2 - x^2}$ (Pythagorean theorem), and thus has an area of $\pi(r^2 - x^2)$.

Let's now look at similar triangles KTC and EGC. Their ratio of similitude is determined by the placement of parallel planes separated at distances r and x.

That is, $\dfrac{NQ}{NP} = \dfrac{r+x}{r}$, which equals $\dfrac{KT}{EG}$,

so that $\dfrac{r+x}{r} = \dfrac{u}{r\sqrt{\pi}}.^3$ \hfill (1)

Similarly, the ratio of similitude of triangles AKL and AEF is $\dfrac{r-x}{r}$,

so that $\dfrac{KL}{EF} = \dfrac{r-x}{r}$, or $\dfrac{v}{r\sqrt{\pi}} = \dfrac{r-x}{r}$. \hfill (2)

Multiplying equations (1) and (2) we get

$\dfrac{uv}{r^2\pi} = \dfrac{(r+x)(r-x)}{r^2}$, or the area of the rectangle $LSTK$ is

$uv = \pi(r+x)(r-x) = \pi(r^2 - x^2)$, which is the area of the circle.

Thus the area of circle H and the area of rectangle $LSTK$ are equal. So by Cavalieri's theorem, the two volumes must be the same.

3. The plane passing through the center of the sphere bisects all the line segments joining points on the two "polar" planes. Therefore, $MP = NP$.

Appendix B

Ramanujan's Work

In this connection it may be interesting to note the following simple geometrical constructions for π. The first merely gives the ordinary value $\frac{355}{113}$. The second gives the value $\left(9^2 + \frac{19^2}{22}\right)^{\frac{1}{4}}$ mentioned on page 114.

(1) Let AB (fig. B-1) be a diameter of a circle whose center is O. Bisect $\overline{AO}$ at M and trisect $\overline{OB}$ at T.

Draw $\overline{TP}$ perpendicular to $\overline{AB}$ and meeting the circumference at P.

Draw a chord $\overline{BQ}$ equal to $\overline{PT}$ and join $\overline{AQ}$.

Draw $\overline{OS}$ and $\overline{TR}$ parallel to $\overline{BQ}$ and meeting $\overline{AQ}$ at S and R,
    respectively.

Originally published as Srinivasa Ramanujan, "Modular Equations and Approximations to π," *Quarterly Journal of Mathematics* 45 (1914): 350–72. Reprinted in *S. Ramanujan: Collected Papers*, ed. G. H. Hardy, P. V. Seshuaigar, and B. M. Wilson (New York: Chelsea, 1962), pp. 22–39.

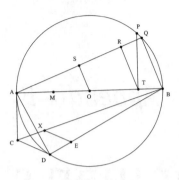

Fig. B-1

Draw a chord $\overline{AD}$ equal to AS and a tangent $\overline{AC}$ equal to RS.

Join $\overline{BC}$, $\overline{BD}$, and $\overline{CD}$; cut off BE equal to BM, and draw $\overline{EX}$, parallel to $\overline{CD}$, meeting $\overline{BC}$ at X.

Then the square on $\overline{BX}$ is very nearly equal to the area of the circle, the error being less than a tenth of an inch when the diameter is 40 miles long.

(2) Let $\overline{AB}$ (fig. B-2) be a diameter of a circle whose center is O.

Bisect the arc $\overparen{ACB}$ at C and trisect $\overline{AO}$ at T.

Join $\overline{BC}$ and cut off from it $\overline{CM}$ and $\overline{MN}$ equal to AT.

Join $\overline{AM}$ and $\overline{AN}$ and cut off from the latter $\overline{AP}$ equal to AM.

Through P draw $\overline{PQ}$ parallel to $\overline{MN}$ and meeting $\overline{AM}$ at Q.

Join $\overline{OQ}$ and through T draw $\overline{TR}$, parallel to $\overline{OQ}$, and meeting $\overline{AQ}$ at R.

Draw AS perpendicular to $\overline{AO}$ and equal to AR, and join $\overline{OS}$.

Then the mean proportional between *OS* and *OB* will be very nearly equal to a sixth of the circumference, the error being less than a twelfth of an inch when the diameter is eight thousand miles long.

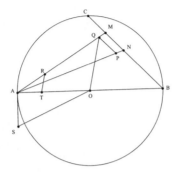

Fig. B-2

Appendix C

Proof That $e^\pi > \pi^e$

From page 146 we provide for the mathematics enthusiast some proofs of the fact that $e^\pi > \pi^e$.

Proof I

$y = f(x) = e^x$ is monotonously growing in **R** (**R** is the set of real numbers)

$x_1 < x_2 \Rightarrow f(x_1) < f(x_2)$

Supposed we know: $e \cdot \ln\pi < \pi$, then we can conclude:

$e \cdot \ln\pi < \pi$ $\Rightarrow$

$f(e \cdot \ln\pi) < f(\pi)$ $\Rightarrow$

$e^{\,e \cdot \ln\pi} < e^\pi$ $\Rightarrow$

$(e^{\ln\pi})^e < e^\pi$ $\Rightarrow$

$\pi^e < e^\pi$

Proof II

$$y = f(x) = x^{\frac{1}{x}} = \sqrt[x]{x}$$

$$y' = f'(x) = x^{\frac{1}{x}-2} \cdot (1 - \ln x)$$

$$y' = 0 \quad \Rightarrow \quad x = e$$

$$f'(e) = -e^{\frac{1}{e}-3} \approx -0.0719... < 0 \quad \Rightarrow \qquad \text{maximum at } x = e$$

$$\text{max}(e; \sqrt[e]{e}) \approx (2.72\ ;\ 1.44)$$

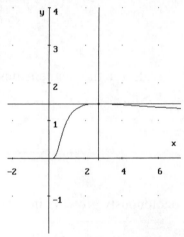

Fig. C-1

maximum at $x = e$ $\qquad \Rightarrow \qquad$ $f(e) > f(\pi)$ $\qquad \Rightarrow$

$$e^{\frac{1}{e}} > \pi^{\frac{1}{\pi}} \Rightarrow (\text{ to the power of } e, \text{ to the power of } \pi) \Rightarrow e^{\pi} > \pi^{e}$$

Proof III

$$y = f(x) = x^{\frac{1}{x}} = \sqrt[x]{x} \qquad\qquad \Rightarrow$$

$$\ln y = \ln x^{\frac{1}{x}} \qquad\qquad \Rightarrow$$

$$\ln y = \frac{1}{x} \ln x \qquad\qquad \Rightarrow$$

$$\frac{y'}{y} = \frac{1 - \ln x}{x^2} \quad = \qquad \Rightarrow$$

$y' = 0:$ Left side $= 0$

     Right side $= 0 \Leftrightarrow$ (numerator) $1 - \ln x = 0 \Rightarrow$

$\ln x = 1$, therefore, $x = e$

$y''(e) < 0 \qquad\qquad\qquad \Rightarrow \qquad$ maximum at $x = e$

and so on as in proof II

Proof IV

For $x > 0$, $e^x = 1 + x + \dfrac{x^2}{2!} + \dfrac{x^3}{3!} + \ldots$, i.e., $e^x > 1$, $e^x > 1 + x$

$\pi > e \Rightarrow \dfrac{\pi}{e} > 1$ and $x = \dfrac{\pi}{e} - 1 > 0$; therefore, $e^{\frac{\pi}{e} - 1} > 1 + (\dfrac{\pi}{e} - 1)$

$1 + x = 1 + (\dfrac{\pi}{e} - 1) = \dfrac{\pi}{e} \Rightarrow$

$\dfrac{e^{(\pi/e)}}{e} > \dfrac{\pi}{e}$, then multiply by e to get

$e^{\frac{\pi}{e}} > \pi$

$e^\pi > \pi^e$

Appendix D

A Rope around the Regular Polygons

We provide here the calculations that enabled us to get the various values for *a* for each of the regular polygons.

For an equilateral triangle:

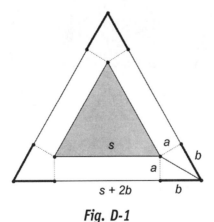

Fig. D-1

The length of the rope is $3s + 1$. The perimeter of the larger triangle is $3(s + 2b) = 3s + 6b$.

With $3s + 1 = 3s + 6b$, it follows immediately that $1 = 6b$ and $b = \frac{1}{6}$.

We know that $\tan 60°$ (or $\tan \frac{\pi}{3}$) $= \frac{b}{a}$, so we get

$$a = \frac{b}{\tan 60°} = \frac{1}{6\sqrt{3}} = \frac{\sqrt{3}}{18} = 0.09622504486\ldots \approx 0.096, \text{ or the length}$$

of a is about 9.6 cm.

For a regular pentagon:

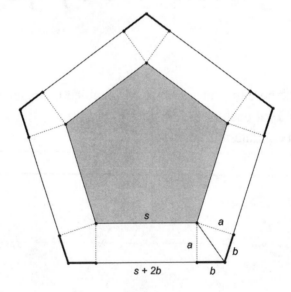

Fig. D-2

The length of the rope is $5s + 1$.

The perimeter of the larger regular pentagon is $5(s + 2b) = 5s + 10b$.

With $5s + 1 = 5s + 10b$, it follows that $1 = 10b$ and $b = \frac{1}{10}$. Since

$\tan 36°$ (or $\tan \frac{\pi}{5}$) $= \frac{b}{a}$, we get

$$a = \frac{b}{\tan 60°} = \sqrt{\frac{\sqrt{5}}{250} + \frac{1}{100}} = 0.1376381920... \approx 0.138,\text{ which indicates}$$

that the distance between the pentagons, a, is about 13.8 cm.

For a regular hexagon:

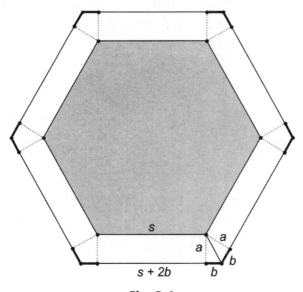

Fig. D-3

The length of the rope is $6s + 1$.
The perimeter of the larger regular pentagon is $6(s + 2b) = 6s + 12b$.
With $6s + 1 = 6s + 12b$, it follows that

$1 = 12b$ and $b = \frac{1}{12}$. Since $\tan 30°$ (or $\tan \frac{\pi}{6}$) $= \frac{b}{a}$, we get

$$a = \frac{b}{\tan 30°} = \frac{3}{12\sqrt{3}} = \frac{\sqrt{3}}{12} = 0.1443375672... \approx 0.144,$$

which indicates that the distance, a, between the hexagons is about 14.4 cm.

The distance a between the respective parallel sides of the rope and initial polygons is also shown by these four regular polygons to be independent of the side lengths of the initial polygons.

For a regular polygon of n sides (called an n-gon:)
The length of the rope is $ns + 1$.
The perimeter of the larger regular polygon is $n(s + 2b) = ns + 2nb$.
With $ns + 1 = ns + 2nb$, it follows that $1 = 2nb$ and $b = \frac{1}{2n}$.
Because $\tan \frac{\pi}{n} = \frac{b}{a}$, we get

$$a = \frac{b}{\tan \dfrac{\pi}{n}} = \frac{1}{2n \tan \dfrac{\pi}{n}}, \text{ or } a = \frac{\cot \dfrac{\pi}{n}}{2n}$$

References

Andersen, David G. "Pi Search." http://www.angio.net/pi/piquery.

Badger, L. "Lazzarini's Lucky Approximation of π." *Mathematics Magazine* 67, no. 2 (1994): 83–91.

Ball, W. W. Rouse, and H. S. M. Coxeter. *Mathematical Recreations and Essays.* 13th ed. New York: Dover, 1987, pp. 55, 274.

Beckmann, Petr. *A History of π.* New York: St. Martin's, 1971.

Berggren, Lennart, Jonathan Borwein, and Peter Borwein. *Pi: A Source Book.* New York: Springer Verlag, 1997.

Blatner, David. *The Joy of π.* New York: Walker, 1997. See http://www.joyofpi.com.

Boyer, Carl B. *A History of Mathematics.* New York: John Wiley & Sons, 1968.

Castellanos, Dario. "The Ubiquitous Pi." *Mathematics Magazine* 61 (1988): 67–98, 148–63.

Dörrie, Heinrich. "Buffon's Needle Problem." *100 Great Problems of*

Elementary Mathematics: Their History and Solutions. New York: Dover, 1965, pp. 73–77.

Eves, H. *An Introduction to the History of Mathematics.* 6th ed. Philadelphia: Saunders, 1990.

Gardner, Martin. "Mathematical Games: Curves of Constant Width." *Scientific American* (February 1963): 148–56.

———. "Memorizing Numbers." *The Scientific American Book of Mathematical Puzzles and Diversions.* New York: Simon and Schuster, 1959, p. 103.

———. "The Transcendental Number Pi." Chap. 8 in *Martin Gardner's New Mathematical Diversions from Scientific American.* New York: Simon and Schuster, 1966, pp. 91–102.

Gridgeman, N. T. "Geometric Probability and the Number π." *Scripta Mathematica* 25 (1960): 183–95.

Hatzipolakis, A. P. "Pi Philology." http://www.cilea.it/~bottoni/www-cilea/F90/piphil.htm.

Havermann, H. "Simple Continued Fraction Expansion of Pi." http://odo.ca/~haha/cfpi.html.

Kaiser, Hans, and Wilfried Nöbauer. *Geschichte der Mathematik.* Vienna: Hölder-Pichler-Tempsky, 1998.

Kanada Laboratory home page. http://www.super-computing.org.

Keith, Michael. *World of Words and Numbers.* http://users.aol.com/s6sj7gt/mikehome.htm.

Olds, C. D. *Continued Fractions.* Washington, DC: Mathematical Association of America, 1963.

Peterson, Ivars "A Passion for Pi." *Mathematical Treks: From Surreal Numbers to Magic Circles.* Washington, DC: Mathematical Association of America, 2001.

Posamentier, Alfred S. *Advanced Euclidian Geometry.* Emeryville, CA: Key College Publishing, 2002.

Rajagopal, C. T., and T.V. Vedamurti Aiyar. "A Hindu Approximation to Pi." *Scripta Mathematica* 18 (1952): 25–30.

Ramanujan, Srinivasa. "Modular Equations and Approximations to π." *Quarterly Journal of Mathematics* 45 (1914): 350–72. Reprinted in *S. Ramanujan: Collected Papers*, ed. G. H. Hardy, P. V. Seshuaigar, and B. M. Wilson, 22–39. New York: Chelsea, 1962.

Roy, R. "The Discovery of the Series Formula for π by Leibniz, Gregory, and Nilakantha." *Mathematics Magazine* 63, no. 5 (1990): 291–306.

Singmaster, David, "The Legal Values of Pi." *Mathematical Intelligencer* 7, no. 2 (1985): 69–72.

Stern, M. D. "A Remarkable Approximation to π." *Mathematical Gazette* 69, no. 449 (1985): 218–19.

Volkov, Alexei. "Calculation of π in Ancient China: From Liu Hui to Zu Chongzhi." *Historia Scientiarum* 2nd ser. 4, no. 2 (1994): 139–57.

Index

313